Development and Characterization of TSP-Filled Basalt and Kenaf Fiber-Reinforced Epoxy Composites

Watkins Angelo

TABLE OF CONTENTS

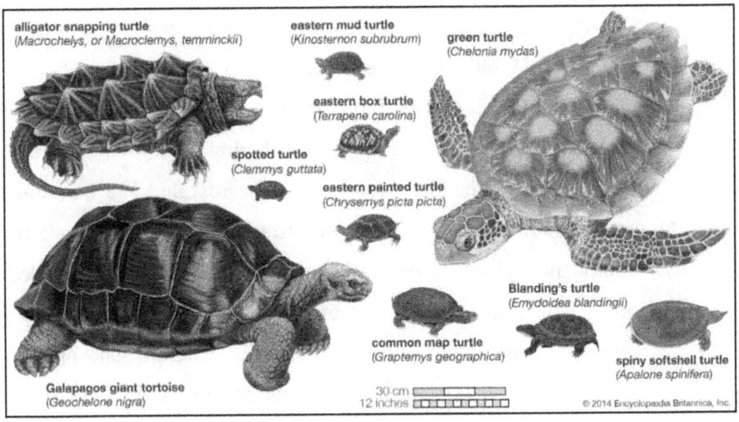

   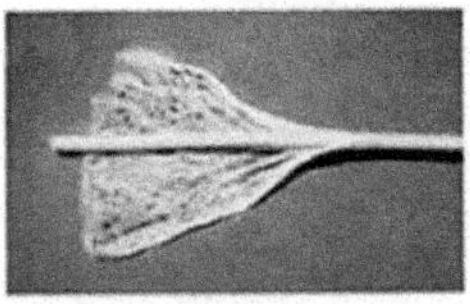 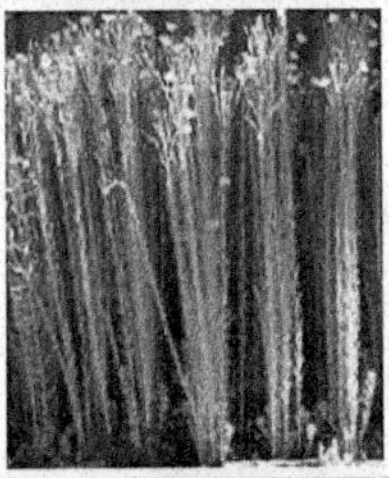 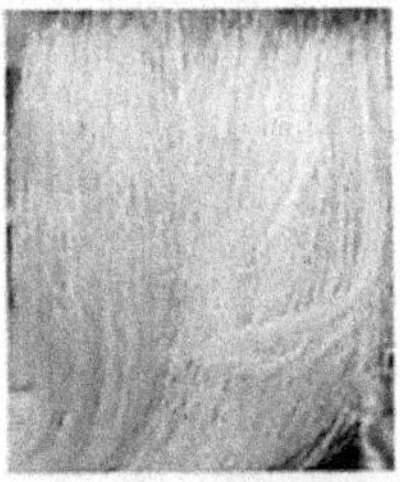  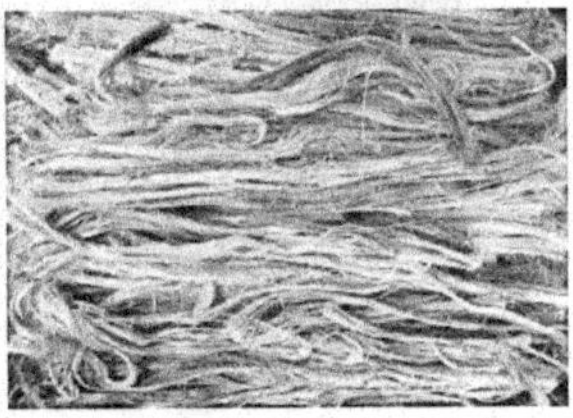     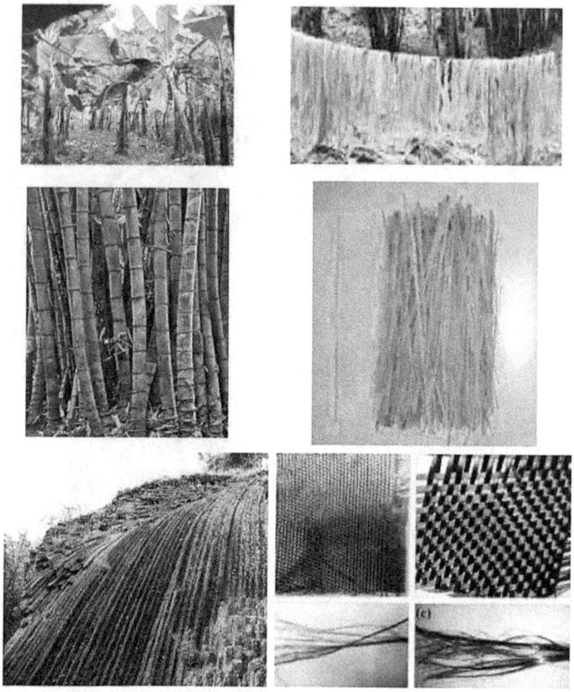

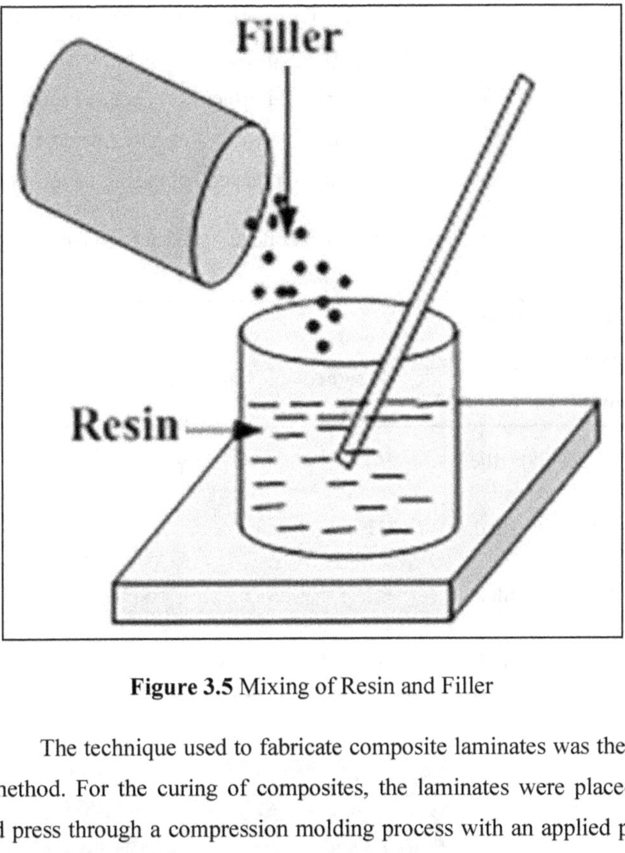

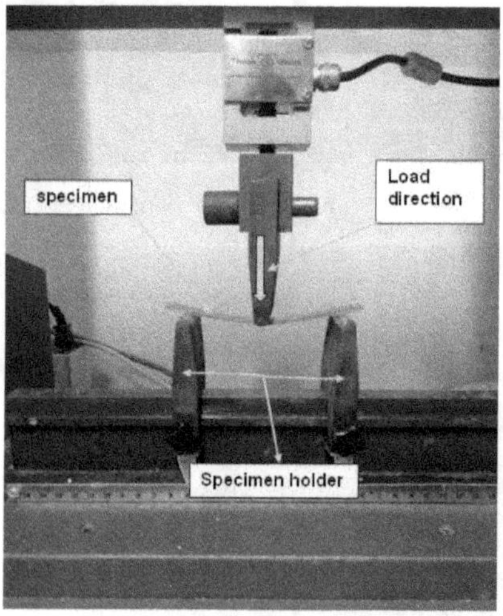

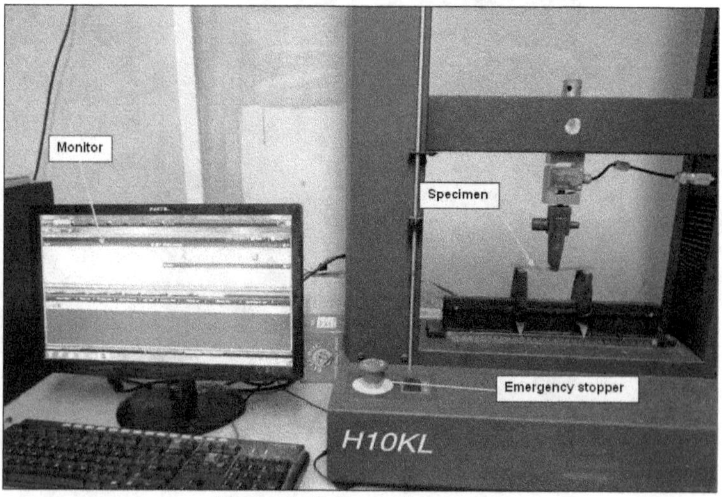

analysis, depicted in Figure 5.8, confirmed the presence of elemental constituents from TSP filler added in kenaf/epoxy composite on the fractured surface of the K5 tensile sample.

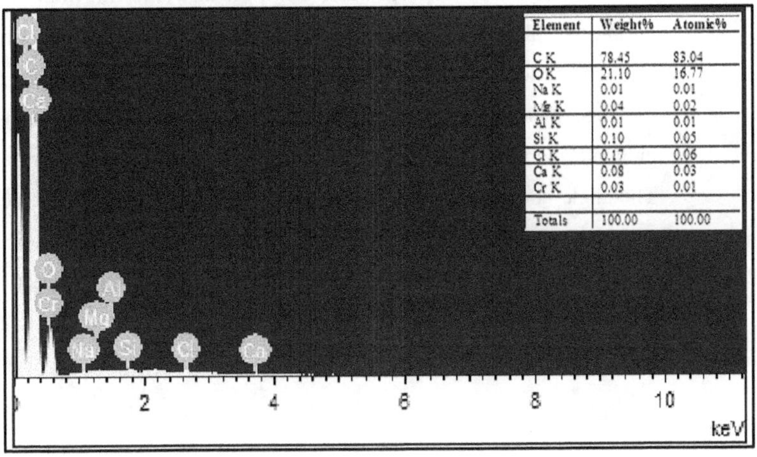

Element	Weight%	Atomic%
C K	78.45	83.04
O K	21.10	16.77
Na K	0.01	0.01
Mg K	0.04	0.02
Al K	0.01	0.01
Si K	0.10	0.05
Cl K	0.17	0.06
Ca K	0.08	0.03
Cr K	0.03	0.01
Totals	100.00	100.00

Figure 5.8 EDAX results for TSP/kenaf/epoxy composite of K5 sample.

5.8 Effects of filler and fiber orientation

The presence of TSP filler improves the composite's enrichment, indicating the highest strength of the material as supported by experimental findings. The filler material comprises minerals that consist of microlamellas referred to as keratin. When the minerals are paired for flexibility and toughness mechanisms, this strengthens the composite [Chen et al., (2017)]. The filler improves mechanical characteristics by improving fracture propagation resistance due to the dispersion of biomineralized nanocomposite platelets. The effective micro and nanostructure of the turtle shell filler contribute to the remarkable overall properties obtained. By including this filler, the matrix modifies the bulk characteristics of the composites' interfacial interaction and builds up the nanostructure inside the phase.

The longevity of the kenaf FRP composite has increased stability and decreased matrix failure while reinforcing [Malik et al., (2021)]. The impact of ply direction control enhances the inherent properties, reducing the failure of

Book title here paste long title here and keep

By
uri gagarin

TABLE OF CONTENTS

CHAPTER – 1
INTRODUCTION

1.1 PREAMBLE

Nature, it is a grand spectacle, teeming with endless possibilities, whispers secrets that could blossom into a vibrant array of scientific achievements. Humanity's early investigation to developed into a perception of dominance, leading to its exploitation and the subsequent crises of climate change, biodiversity erosion, pollution, and resource depletion. To heal our earth's damaged beauty, or at least protect its remaining natural wonders, humanity needs to embrace eco-conscious solutions that simultaneously consider the well-being of nature and the advancement of technology.

1.2 BACKGROUND

The advancements in materials show the different phases of human history, including the Stone Age, Bronze Age and Iron Age. This evidence indicates that materials have played a significant role in human society since its early stages. Composite materials are not a recent development in human history: their origins can be traced back to 1500 BC with the use of such materials by the Mesopotamians and Egyptians. The presence of composite systems is identifiable in vernacular architecture, exemplified by mud-straw bricks, characterized by a dispersed reinforcement phase within a ductile matrix, and wood-clay constructions, demonstrating a combination of structural members and inorganic binder. The Mongol people produced composite bows in approximately 1200AD, constructing them from bone, wood, and animal- based adhesive.

A key turning point for composite materials, occurred in the early 1900s with the advent of plastics. To make newly developed plastic stronger, fibers are added to them. The fusion of fiber glass with a plastic polymer produced a strong yet Leight weight structure. In 1935, Owens Corning developed glass fiber, and its incorporation as a reinforcement led to the creation of the first-fiber reinforced polymer. Even before World War II finished, a small industry focused on specialized composite materials was starting to grow. Subsequently, a decreased focus on military utilization, several pioneering researchers in composite technology were actively engaged in efforts to incorporate these heterogeneous materials into diverse industrial applications. In 1946, the initial commercial watercraft body made from composite materials was lunched and the growth of composite industries were started in 1970s.

1.3 MOTIVATION

The necessity for novel material systems characterized by improved performance characteristics has become considerably more pronounced in order to satisfy the evolving demands of industrial processes. The utilization of composite materials systems offers a potential solution to address the stipulated performance criteria. Properties modulation to align with specified performance parameters is achievable through the controlled variation of the matrix and reinforcement components with in the composite architecture. The design and fabrication of polymer composite material is a platform to enable the attainment of materials exhibiting both low areal weight and enhanced superior mechanical strength.

Over the past few decades, the scientific community has shown a notably heightened interest in pursuing the development of polymer composites. Numerous technological advancements have occurred in the development of novel processing routes and the evolution of advanced reinforcements fibers which citing aramid, carbon, basalt, kenaf, jute and ramie fibers etc. as prominent examples. Recently, the incorporation of filler as additional material, to enhance the high-performance strategy of polymer matrix composites has been a key factor in their growing demand. The primary impetus for the expansion of natural fillers within fiber-reinforced composites is the increasing need for lightweight, recyclable, and eco- compatible benign products.

From materials science standpoint, the advancement of composites has yielded a substantial increase in the achievable diversity of property sets. Present research focuses on the continued optimization of existing composites and the creation of entirely new material systems to access even wider property domains.

1.4 COMPOSITE MATERIALS

Composite materials arise from the blending of two or more physically and chemically diverse substances, leading to a material with properties that exceed those of its separate parts. The structure of polymer composites involves a polymer as the matrix, with fibers serving as reinforcement, which exhibits both chemical and physical attributes. Besides the matrix and reinforcement components, the strength of polymer composites is also increased by the inclusion of additives and fillers within the matrix. While these components are joined, the individual constituents preserve their unique identities and properties. Polymer composites are increasingly recognized as a sophisticated category of engineering materials owing to their stable dimension combined with a high strength-to-weight ratio, specific mechanical characteristics, and favourable resistance to corrosion and thermal environments. These materials utilized in diverse sectors such as aerospace, defence, marine, civil engineering, structural applications, the automotive industry, electrical engineering and even sports equipment.

1.5 COMPOSITE TYPES

The fundamental makeup of composite material's structure involves two distinct components: a primary phase and a secondary phase. Characteristically, the secondary phase in a composite is discontinuous, possesses greater hardness and strength than the continuous primary phase (the matrix), and functions as the reinforcement. When forming a two-components composite, the matrix and the reinforcement can each be selected from the material classes of metals, ceramics or polymers, resulting in the possible groupings of metal-metal, metal-polymer, ceramic-ceramic, ceramic-polymer, and polymer-polymer. The essential purposes of the matrix phase in a composite are to: (i) give the material its overall form, (ii) keep the reinforcement properly oriented and (iii) contributed to the material's strength by sharing applied forces with the reinforcement. The morphology of the reinforcing phase in a composite can vary widely, including fibers, particles, short single crystals (whiskers), thin planar shapes (flakes), and a range of other geometric arrangements.

1.5.1 Composite – Metal Matrix

Currently metal matrix composites (MMCs) are attracting significant research interest, yet they aren't as commonly used as polymer composites, despite MMCs offering superior properties like high strength to weight ratio, better wear resistance, good thermal conductivity, good electrical conductivity, and better strength at elevated temperature [Dey et al., (2015), Gecu et al.,

(2019), Venkatesan et al., (2017)].The superior properties of MMCs are directly attributable to the inherent qualities of the matrix and the reinforcing phase. The quiet common materials utilized include aluminium, magnesium, copper, titanium, superalloys, zinc and their respective alloys. MMCs present certain drawbacks compared to polymer matrix composites, notably the greater expense of some constituent material systems and the complexity of ensuring a uniform dispersion of the reinforcement.

1.5.2 Composite – Ceramic Matrix

The defining characteristic of ceramics as solid materials is their exceptionally strong ionic bonding, occasionally supplemented by covalent bonding. CMCs development has trailed other composites, stemming from two major contributing factors. The initial impediment is that many processing pathways for CMCs demand high temperatures, consequently requiring the exclusive use of high-temperature-resistant reinforcements. The use of high temperatures in the production of CMCs is the second factor that has hindered their progress. When the matrix and reinforcement cool, the disparities in their thermal expansion coefficients cause internal thermal stresses. Due to their demanding manufacturing processes and high costs, metal and ceramic matrix composites, even the most advanced ones, have fewer applications than polymer matrix composites. The desirable properties of CMCs, including high strength, hardness, high-temperature resilience, chemical inertness and low density, make them highly advantageous materials [Ilyas et al., (2021), Zhang et al., (2012)]. The disadvantage of CMCs is their catastrophic failure under tensile or impact forces, which stems from their inherently low fracture toughness. The ceramic matrix materials include Silicon Carbide(SiC), Alumina (Al_2O_3), Boron Carbide (B4C), Boron Nitride (BN), Aluminium Nitride (AIN), and Titanium Carbide (TiC).

1.5.3 Composite – Polymer Matrix

Polymer Matrix Composites (PMCs) represent a subset of composite materials identified by the use of a polymer as their matrix. A standard definition of polymers is that they are extensive molecular chains made up of many identically arranged repeating units. The fundamental architecture of polymer bonding manifests as linear, branched, cross-linked or network structures. A polymer consists of large molecules formed from numerous identical repeating units. PMCs have two main constituents: a polymer serving as the matrix and fibers acting as the reinforcement. Generally, the fibers used in composites possess superior strength and modulus compared the polymer matrix, enable them to function as the main load-bearing elements of the structures. The matrix material's function is to ensure strong adhesion with the fibers, enabling it to effectively transfers load.

The characteristics of a composite material are closely tied to the individual properties of its polymer matrix and fiber reinforcement, as well as how these two components interact. The market offers diverse types of composite materials selected based on the applications and each possessing distinct performance characteristics. Common methods for fabricating PMCs include hand lay-up, filament winding, resin transfer moulding, pultrusion, compression molding and vacuum impregnation. The advantages of polymer composites are lightweight, high strength, strength-to-weight ratio, corrosion resistance, High-Impact strength, design flexibility, low thermal conductivity, etc. [Keya et al., (2019), Oladele et al., (2020), Asim et al., (2018)].

1.5.4 Composite – Hybrid

A super material where a foundational matrix embraces strong fibers for backbone and then, as a bonus, tiny particulates are sprinkled in, creating a true hybrid composite. The intent of integrating fiber and fillers within the matrix is to create a hybrid composite that ideally demonstrates the favourable attributes of the individual composites while concurrently reducing certain undesirable properties [Thombre et al., (2014)]. The combined use of particles and fibers in polymer matrices, which could lead to significantly improved properties and wear performance through synergy, has not been thoroughly researched. Howbeit, recent reports indicate that the incorporation of filler particles into the matrix of fiber reinforced composites may yield synergistic effects, resulting in an increased modulus and decreased material cost. In the evaluation of hybrid composites, material scientists can readily recognize multiple key benefits. One significant advantage of hybrid composites is the ability to manage costs by substituting expensive fibers or matrix materials with more affordable fillers or alternative fibers. A second advantage lies in the ability to broaden the composite's mechanical and physical properties spectrum through the strategic selection of reinforcements with varying attributes. Finally, its's possible to develop novel properties in hybrid composites that cannot be realized in single-reinforcement composite systems. This approach offers significant flexibility in developing advanced materials, allowing us to achieve desired properties by harnessing the combined influence of multiple reinforcement particles [Subramanian et al., (2015), Geng et al., (2006)]. The field of composites incorporating multiple reinforcements is a contemporary focus of research, primarily because of their superior performance attributed to upgraded mechanical characteristics and tribological response.

1.6 CLASSIFICATION OF POLYMER COMPOSITES

Polymer composites predominantly utilize either thermoset or thermoplastic polymers as their matrix materials. Elastomers also serve as a additional type of matrix employed in composite materials. Again, polymer composites are categorized into particulate-filled, fiber-filled, and network composites, based on how the reinforcement is incorporated. The overall classification of the polymer composites is illustrated in figure1.1.

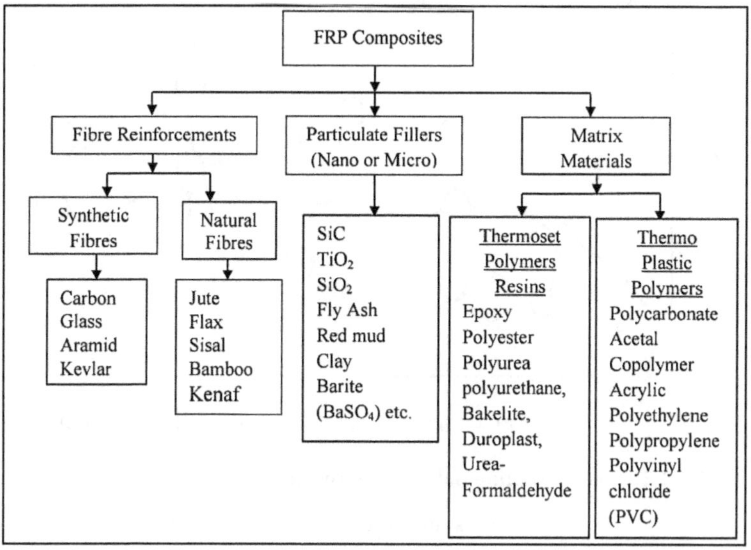

Figure 1.1 Classification of Polymer Composites.

1.6.1 Based on Matrix Phase Classification

In its context of composite materials, the matrix is the continuous phase that surrounds and binds together the reinforcement by its characteristics of cohesive and adhesive values. Its's crucial because it transfers the load to the reinforcement, protects the reinforcement from environmental damage, and provides the overall shape and integrity of the composite.

1.6.1.1 Thermoplastic polymer composites

Thermoplastics are materials that become pliable when heated and solidify into a fixed shape upon cooling. These polymers could soften multiple times when reheated. The morphology of these materials, determined by their molecular structure, can be crystalline or amorphous, which directly impacts their diverse softening temperatures. During heating process, you will need to be careful to prevent the plastic from degrading or decomposing. It is typically easier to process thermoplastics and shape them into complex designs compared to the thermoset plastics. **Figure1.2** shows the molecular structure of the thermoplastic. The most commonly used thermoplastics are Low-density Polythene (LPDE), High-density Polythene (HDPE), Polytetrafluoroethylene, Polystyrene, Polypropylene, Polyvinyl Chloride (PVC), Polymethyl Methacrylate (PMMA, acrylic) and Polyamide (nylon).

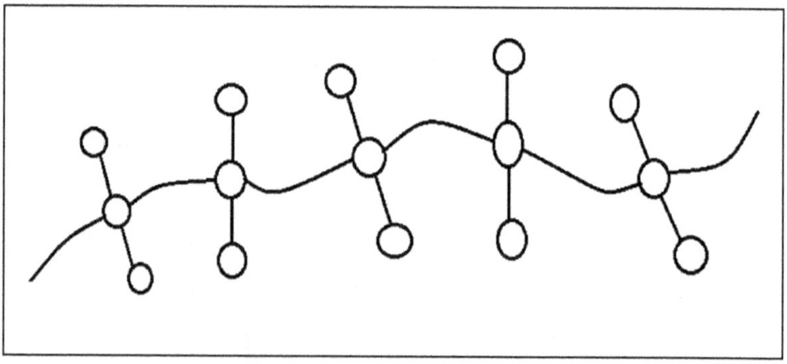

Figure1.2 Thermoplastic – long chain polymer molecule.

1.6.1.2 Thermosetting polymer composites

Thermosets are materials produced through the chemical bonding of macromolecules, forming a strong, extensively cross-linked polymer network.

In essence, they consist of large molecules tightly interconnected by chemical bonds, leading to a durable and rigid structure. The use of thermosetting polymers as a matrix for high-performance naval and aerospace structures is now a well-established practice. These polymers are typically synthesized and moulded simultaneously. When heated, they may soften slightly but cannot be reshaped and remoulded significantly, as they do not flow like thermoplastics. It is the chemically formed, highly crosslinked structure of thermosets that gives them an edge in mechanical and physical properties over thermoplastics and elastomeric counterparts. On the flip side, the extensive crosslinked network in these materials leads to drawbacks like brittleness, insolubility, creep behavior, and limited flame resistance and elasticity. To overcome these drawbacks, extensive research has focused on modifying these polymers with reinforcing organic and inorganic fillers. Figure1.3 shows the molecular structure of the thermoset polymers. Today's industrial applications often feature thermosetting polymers such as polyamides, polyimide, epoxies, vinyl ester and phenolic resins.

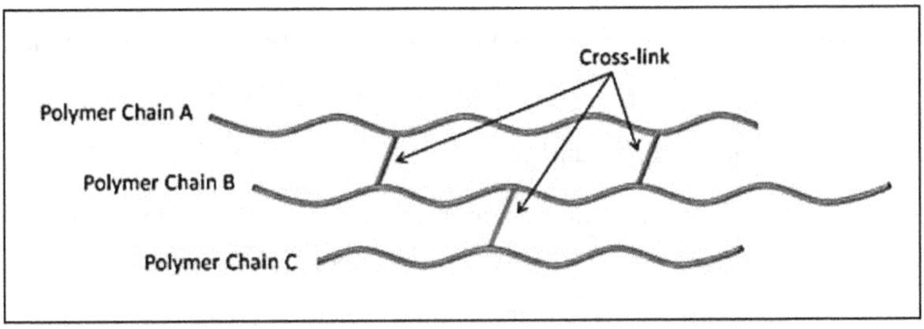

Figure1.3 Thermoset- cross link polymer molecule

1.6.1.3 Elastomer Composites

Elastomers are a class of polymers characterized by their viscoelastic properties, stemming from their composition of long, chain-like molecules. Once the applied force is released, the elastomer will recover its original configuration. Elastomers can be divided into two main categories: saturated and unsaturated. Application like seals, bushings, and adhesives frequently utilize elastomers.

1.6.2 Based on Reinforcement Phase Classification

The primary function of the reinforcement in composite materials is to enhance the mechanical properties of the pure resin. The specific properties of the composite are influenced by the type of filler used, as different fillers impart distinct characteristics. While individual fibers or bundles are suitable for certain processes like filament winding or pultrusion, most applications require fibers to be arranged into sheets or other forms for easier handling. This arrangement of fibers into various orientations results in different types of fabrics, each with unique characteristics. Ultimately, composites reinforced with fabric exhibit significantly superior mechanical properties compared to unreinforced resin systems [Aabdul et al., (2009)]. In fabric-reinforced composites, the mechanical performance is largely governed by the properties of the reinforcing fibers. The following four primary aspects govern the role of fibers in a composite:

- The fiber's own mechanical characteristics.
- The bonding quality at the fiber-resin interface.
- The proportion of fibers present (volume fraction).
- The directional alignment of the fibers.

1.6.2.1 Fiber Reinforcement

Fiber reinforcement in composites is achieved using continuous (long) or discontinuous (short) fibers embedded in matrix. Their anisotropic characteristics make them highly efficient at increasing fracture strength. The orientation of molecular structures in polymers determines their mechanical strength. Fibers, due to their limited cross-sectional size, are not suitable for direct use and must be integrated into a polymer matrix to create fibrous composites for engineering purposes. Reinforced composites are produced in multiple shapes such as sheets, tapes, and related formats and are typically manufactured through thermosetting methods [Mitschang et al., (2003)]. The fiber reinforcement is discussed in detail in the subsection of 1.7

1.6.2.2 Particle Reinforcement

Particulate composites are a type of composite material formed by embedding small particles of one or more substances within a polymer matrix, which can be either thermoplastic or thermoset. These particulates can vary widely, including fine particles, chopped fibers (like glass fibers), platelets, hollow or solid spheres, silica, or alumina. Such composites significantly enhance the mechanical, electrical, flame retardant and barrier properties of polymers. Typically, these new materials are incorporated into the base polymer in small amounts, and their mixing process is consistent with that used for other particulate composites.

A key challenge in particulate composites is the agglomeration of filler particles, which often results in mechanical properties falling short of expectations. In polymer composites with dispersion reinforcement, the particles are remarkably small, typically between 0.01 to 0.1 mm in diameter. This strengthening mechanism operates at an atomic or molecular level, akin to precipitation hardening in metals. Here, the polymer matrix primarily carries the applied load, while the dispersed particles effectively obstruct or slow down the movement of dislocations, thereby enhancing the material's strength [Chawla et al., (1998)]. A broad spectrum of applications for particle-reinforced composite materials includes, but is not limited to, the automotive sector, household appliances, electrical products, toys, casings for mobile phones and computers, office furniture and helmets.

1.6.2.3 Flake Reinforcement

Often, flakes are reinforced within a polymer matrix by laying them out in a two-dimensional arrangement, creating a strong bond between the flakes and the polymer. Likewise, composite materials can be engineered with flakes and ribbons as reinforcements, with these elements arranged parallel to each other in two dimensions. In terms of performance, this composite material surpasses fiber- reinforced polymer composites. Conversely, planar reinforcement in composites is a relatively nascent idea. The scarcity of planar reinforcements, unlike their widely available fiber counterparts, has consequently resulted in less research and development in this area [Rexer et al., (1979)].

1.6.2.4 Filler Reinforcement

Fillers are substances incorporated into the polymer matrix to enhance its properties. They can appear in various forms, such as particles or short fibers (whiskers), with particle fillers being more commonly used due to their ease of dispersion. The choice of filler largely depends on the type of matrix employed. Moreover, filler selection largely depends on their inherent characteristics. Important considerations include size, shape, distribution, and specific surface area. The core purpose of adding fillers is to improve the composite's properties while minimizing production expenses. The addition of fillers to the matrix phase provides a way to restructure and fine-tuning of various properties, including mechanical, thermal, tribological, optical and electrical performance.

1.7 Fiber Reinforced Polymer composites

The demand for lightweight, strong, and rigid constructions often leads to the use of fiber-reinforced polymer composites, which excel at improving strength-to-weight ratio and stiffness-to-weight ratios. Most fiber-reinforced composites are created by embedding glass or metal fibers within a polymer matrix. They are essentially materials made of a polymer matrix bolstered by high-modulus fibers. In a composite structure, the fibers are the key elements for carrying loads. They are surrounded by or connected to a matrix, which not only facilitates the transfer of these loads but also offers protection from environmental damage, such as that caused by extreme heat and moisture [Mallick et al., (2007)]. The advantages of FRP composites are as follows.

- High strength -to-weight ratio.
- High stiffness-to-weight ratio.
- Corrosion, Fatigue and Impact Resistance.
- Design Flexibility.
- Thermal and Electrical Insulation.
- Dimensional stability.
- Durability and long service life.
- Minimal cost and Low maintenance.
- Easily recyclable.

The dominant key factors related to fibers that influence the mechanical performance of FRP composites are

- **Fiber Length:** fibers used in composites may be either short or long, with continuous long fibers typically being favoured. Whilethe simplicity in processing and affordability of short fiber composites make them highly suitable for industrial-scale manufacturing operations. It can be organized in randomly or in sequence order structures. In the case of long fiber composites, the fiber withstands the applied load due to stiffness and strength of the fiber when integrated with matrix material and hence, it is suitable for engineering applications. Long fibers offer several benefits, including improved impact resistance, lower shrinkage, enhanced durability, superior surface finish and better dimensional stability (Kaw 2009). Conversely, short fibers are cost-effective, easy to handle, durable, allow for faster production cycle, exhibit fewer defects and can provide higher strength.
- **Fiber Orientation:** When fibers are arranged in a single direction, they confer substantial stiffness and strength, but along that specific axis. For high stiffness and strength, it is necessary to align the fibers in various directions.
- **Fiber Shape:** It is in the context of composite materials, refers to the cross-sectional geometry of the individual reinforcing fibers. While the most encountered fiber shape, especially in glass fibers, is circular, various other shapes exist and increasingly explored to use, particularly in natural fibers and some carbon fibers. With advancements in fiber production, non-circular fibers are gaining prominence, especially in carbon and natural fibers, and more recently, in engineered glass fibers.
- **Fiber Material:** The type of fiber material significantly influences the mechanical properties of a composite. Fibers are expected to possess high elastic modulus and strength, which is why materials like graphite, aramid and glass dominate the fiber composite industry – balancing both performance and cost. Based on the

desired engineering applications both synthetic and natural fibers are used.

Composites are extensively used in diverse fields, with hundreds of variants supporting everything from structural elements in buildings to automotive and aerospace parts. The widespread use of FRP across industries is attributed to the multiple benefits it offers in terms of performance, strength and design flexibility. Thanks to its broad range of benefits by FRP composites.

1.7.1 Natural FRP composites

Natural FRP composites utilize sustainable plant-based fibers embedded within a polymer matrix. Common natural fibers flax, hemp, jute etc., which can be combined with both thermoset and thermoplastic resins. These composites offer several advantages, such as being lightweight, renewable, recyclable, biodegradable and having a lower environmental impact compared to synthetic FRPs [Dwivedi et al., (2015)]. They are also generally more cost-effective. However, natural FRPS can suffer from higher moisture absorption, leading to reduced mechanical properties and durability and possess lower absolute strength and stiffness compared to synthetic fibers like carbon and glass. Despite these limitations, they are gaining traction in various applications, including automotive interiors, packaging, furniture and some consumer goods as industries used. These composites support circular economy goals by reducing carbon footprints. Overall, natural FRP composites represent a promising step toward greener material technologies.

1.7.2 Synthetic FRP composites

Synthetic Fiber Reinforced Polymer (FRP) composites are advanced materials where high-performance synthetic fibers are embedded within a polymer matrix to achieve superior mechanical properties. The most common synthetic fibers include glass, carbon, and aramid, each offering distinct advantages like cost-effectiveness (glass), exceptional strength and stiffness (carbon), or impact resistance and toughness (aramid). These fibers are bound by either thermosetting polymers (e.g., epoxy, polyester) for rigid, strong structures or thermoplastics (e.g., nylon, polypropylene) for better impact resistance and recyclability. The primary benefits of synthetic FRPs are their outstanding strength-to-weight and stiffness-to-weight ratios, excellent corrosion and fatigue resistance, and the ability to tailor properties through fiber orientation. Consequently, they are extensively utilized in demanding applications across aerospace, automotive, marine, wind energy, and civil infrastructure, driving innovations that require lightweight, high-performance, and durable structural solutions [Asyraf et al., (2022), Rajak et al., (2022)].

1.8 Natural Fibers

For more than three millennia, natural strands – fibers sourced from plants and other organisms – have been utilized for their strengthening properties. Based on their botanical origins, these natural fibers fall into three main classifications. They are namely, plant, animal and mineral fibers. Natural fibers, frequently derived from plants, come in five distinct types: leaf, fruit, stalk, bast, and seed fibers. The composition of these plant – based fibers largely include cellulose, hemicellulose and lignin. The structure of fibers is strengthened, stiffened and stabilized by cellulose, a non-branched macromolecule composed of a chain of 1-4 linked β-d-anhydrooglucopyranose units of variable length. As the second most prevalent family of naturally occurring polymers, hemicellulose is an essential part of natural cellulosic fibers. Notably, lignin, a highly branched polymer, works with hemicellulose to form a matrix that embeds the cellulose fibers in the cell wall; as identified in figure1.4.

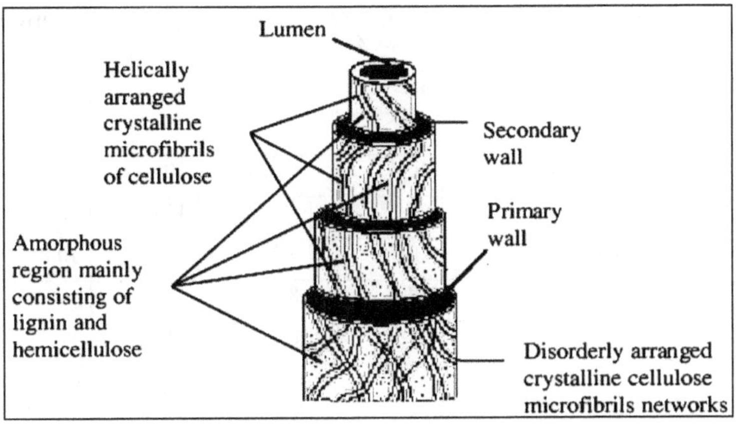

Figure 1.4 Schematic representation of structure of natural fiber [Kalia et al., 2009].

Several key factors influence the properties of natural fibers, including their porosity, swelling behavior, wettability, moisture absorption, geometric attributes (diameter, length, and shape), fiber orientation, renewability, the reliability of their supply, chain, and their chemical makeup. Tables 1.1 & 1.2illustrate the mechanical and chemical composition of disparate natural fibers.

Table 1.1 Typical properties of natural fibers [Sengupta et al. 2020]

Fiber	Density (g/cc)	Tensile Strength (MPa)	Elastic Modulus (GPa)	Elongation (%)
Coir	1.5	593	4-6	30
Flax	1.5	500-1500	27.6	2.7-3.2
Kenaf	1.45	930	53	1.6
Jute	1.3	393-770	26.5	1.5-1.8
Hemp	1.47	690	70	2-40
Sisal	1.5	511-635	9.4-22	2.0-2.5
Bamboo	0.6-1.1	140-230	11-17	
Softwood kraft Pulp	1.5	1000	40	4.4

Table 1.2 Chemical Composition of Natural Fibers [Fan and Fu, 2017]

Fibers	Chemical components				
	Cellulose (wt%)	Lignin (wt%)	Hemi cellulose (wt%)	Moisture (wt%)	Wax (wt%)
Coir	43	45	0.3	0	-
Flax	85	4	9	10	1.7
Bamboo	26-43	21-31	30	-	-
Jute	61	12	13.6	12.6	0.5
Hemp	70.2	3.7	17.9	10.8	0.8
Sisal	78	8	19	11	2
Cotton	85-90	0.7-1.6	01-03	1	-
Wheat Straw	51	16	26	-	-
Aloe vera	67.4	13.7	8.2	5.8	0.24

Sansevieria ehrengergii	80	7.8	11.25	10.55	0.45
Cocconiagrandis.L	62.35	15.61	13.42	5.6	0.79

1.9 Fillers

The alternative decision to add fillers to a polymer is drives by two main objectives: improving its mechanical and thermal characteristics of reducing manufacturing expenses. Fillers are generally classified into organic and inorganic fillers. These fillers play a crucial role in enhancing the performance of FRP composites. Organic fillers, such as natural fibers (e.g., Kenaf, jute, coir and hemp) and agriculture byproducts are lightweight, biodegradable and cost effective. They improve specific strength, reduce density and make composites more eco-friendly, though they may have limitations in thermal stability and moisture resistance. Inorganic fillers like silica, alumina, calcium carbonate, glass beads and metal oxides offer superior mechanical strength, dimensional stability, thermal resistance and wear resistance. They are often used to improve stiffness, hardness and durability but may increase the composites weight and cost. By selecting the right combination of organic and inorganic fillers, FRP composites can be tailored to meet the mechanical, thermal and environmental requirements of various engineering and industrial applications.

The primary reason particle fillers are often chosen is their ease of dispersion. The selection of a suitable filler should be done with care, considering both the matrix type and the specific properties of the filler, which are key determinants. Among the various factors influencing filler choice, filler size, shape, distribution and specific surface area stand out as critical considerations. Essentially, fillers are incorporated to enhance both material's properties and lower its cost. When properly dispersed in a matrix, they can significantly restructure the material's mechanical, thermal, tribological, optical and electrical performance. Normally, particulate fillers are come in two size ranges either at the micron level(under 100 microns) or the nano level (1 to 10 nanometers). The particle's shape is a crucial factor, influencing composite characteristics such as tensile properties, hardness, stiffness, impact behavior and the smoothness of its surface. The distribution of filler particles within a composite is influenced by the filler's inherent characteristics. Depending on their nature, these dispersions can lead to the creation of distinct particles, fragments. aggregates or agglomerates. Figure 1.5 depicted the different form of filler material.

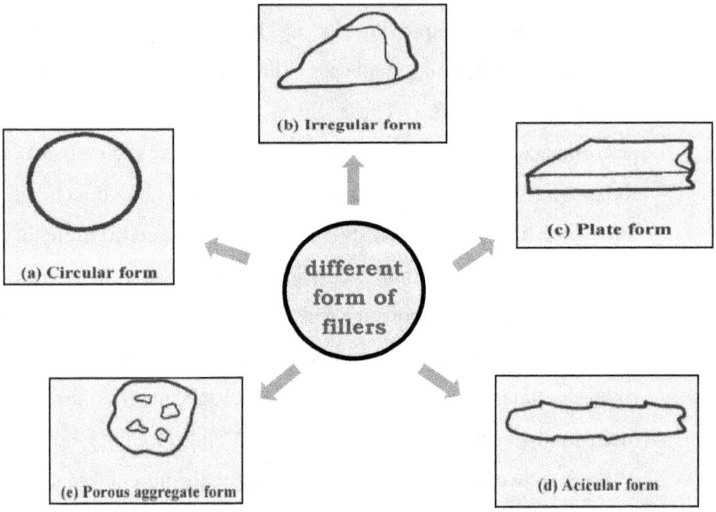

Figure 1.5 Various formation of filler materials [Rothon, 2003]

1.10 Fabrication techniques of FRP composites

The fabrication of polymer composite materials reinforced with Is done using different processing techniques, which are divided into three main categories: Manual automatic and semi-automatic. Some of these techniques include

- **Hand Lay up:** A manual process where layers of fiber reinforcement are placed in a mould and then saturated with resin by hand.
- **Filament winding:** An automatic process where continuous fiber strands are wound around rotating mandrel.
- **Pultrusion:** An automatic process that pulls continuous fibers through a resin bath and a heated die to create a profile with a constant cross-section.

- **Resin transfer moulding (RTM):** An automatic process where dry fibers are placed in a closed mould and then liquid resin is injected under pressure.
- **Resin infusion moulding:** Similar to RTM, but the resin is drawn into the mould using a vacuum instead of being injected.
- **Compression moulding:** The process where a pre measured amount of composite material is placed into a heater mould and then pressed into a shape.
- **Centrifugation:** A process that uses centrifugal force to distribute resin and fibers evenly within a rotating mould.
- **Vacuum assisted resin transfer moulding (VARTM):** A vacuum is used to draw resin into a dry fiber perform that has been sealed between a rigid mould and a flexible vacuum bag.
- **Vacuum infusion:** This process uses vacuum pressure to draw Liquid resin Into a dry fiber material that has been sealed inside a mould.

1.10.1 Hand layup process

Among all fabrication methods, hand layup is considered the most straightforward and basic technic for producing fiber reinforced polymer composites.

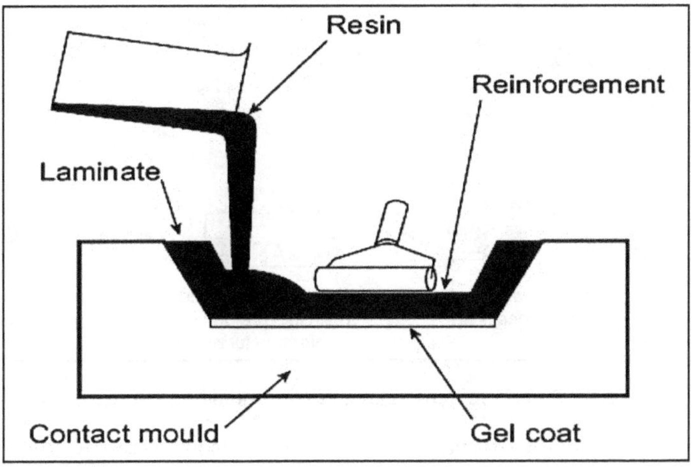

Figure.1.6 Hand lay-up process [Irving &Soutis, 2020].

The key benefit of this process is its simplicity, as it avoids the need for specialized equipment. It is applicable for both limited and large-scale production, offering flexibility at a relatively low cost. The process illustrated in the figure1.6 starts with the application of a gel coat on the mould, followed by the placement of fiber reinforcement. Resin is rolled over the fibers, promoting proper wetting, laminate consolidation and the removal of air bubbles.

1.10.2 Compression Moulding

Compression moulding is a high-volume high-pressure manufacturing process ideal for complex fiber reinforced polymer FRP components which is shown in figure 1.7. The process uses large hydraulic or mechanical pressures with Preheater Metal moulds 250 slash 40°F. A resin charge is placed inside a stationary mould and a movable, heated mould Of is used to place the material into the desired shape. The process has a short cycle time and can be used with various materials, including sheet moulding compounds (SMC), Bulk moulding compounds (BMC), thick moulding compounds (TMC) and wet layup compression moulding. It is also common to use this method in conjunction with the hand layup technique For FRP fabrication.

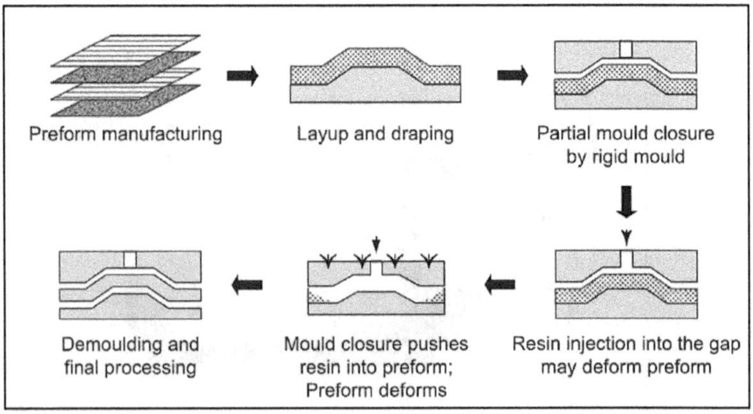

Figure 1.7 Procedure steps of compression moulding process [Irving &Soutis, 2020]

It is appreciated for its user-friendly process, cost-effectiveness, and efficient use of materials with minimal wastage. Despite these benefits, it faces limitations since the method is time-consuming, making it more suitable for small-scale manufacturing. Moreover, the technology demands experienced and highly skilled personnel to operate and monitor the procedures correctly [Rong et al., (2001)].

1.11 Epoxy Resin

First commercialised in the 1940s epoxy resin Is now a staple In engineering due to its superior properties. It's known for its low shrinkage, rigidity, and resistance to chemical corrosion. Additionally, it provides excellent adhesion and thermomechanical performance and high dry electric strength. In contrast to polyester resins, epoxies are capable of maintaining reliable mechanical and physical performance even when subjected to adverse operating conditions [Pulikkalparambil et al., (2021)]. Epoxy resin can be used to bond with a wide variety of materials, such as wood, stone, glass, plastics, ceramics and metals [Stenko et al., (2001)]. Epoxy resins are low-weight polymers with epoxide (oxirane) rings in their structure. A wide variety of epoxy are commercially available which includes diglycidyl ether of bisphenol, LY556, cycloaliphatic epoxy, triglycidyl phenol, tetraglycidyl diamino diphenylmethane and novoloc resins [Vijayan et al., (2015), Kausar et al., (2017)].

Epoxy resin in a composite serves a dual purpose: it distributes force evenly to the reinforcing materials and protects the composites structural integrity. Much research has been done on combining epoxy resins with the natural fibers, synthetic fibers and various reinforcing particles [Abdellaoui et al., (2015)]. Common synthetic reinforcements for epoxy composites include glass, carbon, kevlar and aramide fibers. These composites are often used in automotive and construction industries because they're lightweight, strong and have a good modulus. A notable downside, however, is the difficulty of recycling these synthetic fibers after their use. Epoxy is considered an excellent choice of resin for both synthetic and natural fibers The addition of micro or nanofillers to the resin can also alter the composite's mechanical properties.

1.11.1 Diglycidyl ether of bisphenol A

For an epoxy pre-polymer to be converted into a solidified, networked product (an infusible material), its structure requires a minimum of two functional groups. These pre-polymers currently constitute the vast majority of the global epoxy market, making up approximately three-quarters a(75%) of world production. Condensation polymerization is the method used to synthesize Bisphenol A diglycidyl ether resin (DGEBA). This process involves combining the fundamental chemicals Epichlorohydrin (ECH) and Bisphenol-A (BPA). Figure1.8 illustrates the chemical reaction for the formation of epoxy resin.

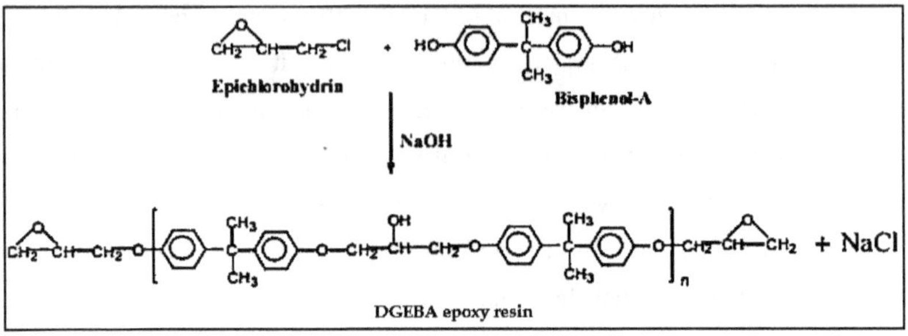

Figure 1.8 Chemical reaction for the formation of epoxy resin

By controlling the reaction between Bisphenol-A and Epichlorohydrin, the molecular weight of epoxide resin can be adjusted. A decrease in epichlorohydrin concentration produces low-molecular-weight epoxy resins that exist as liquids, while higher molecular weight resins are viscous or solid. The symbol n refers to the number of repeating units separating the epoxide rings. The defining characteristic of the wide range of thermosetting matrix resins known as epoxy resins is their epoxide or oxygen functionality (Figure1.9).

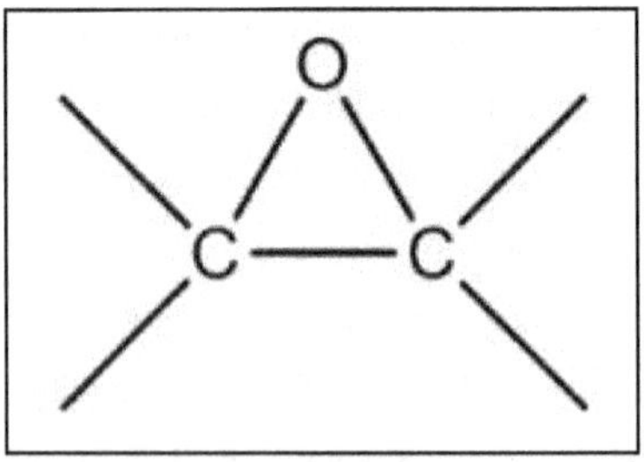

Figure 1.9 Epoxide chemical structure

1.11.2 Applications of epoxy resin

The numerous applications of epoxy resins can be effectively grouped into several major categories based on their function and end-use environment.

Category	Typical Applications
⊙ Protective Coatings & Linings	Underwater ship hulls, Cargo tank linings, Steel bridges, Storage tanks (metal and concrete), Coil coatings for household appliances, Automotive coatings, Food and drink cans/can ends.
⊙ Electrical & Electronics	Printed circuit boards (PCBs), Electrical encapsulation (for motors, engines, machinery).
⊙ Construction & Infrastructure	Construction panels (cladding, metal roofing, ceilings, garage doors), Flooring (industrial/public buildings, food/catering industry, chemical plants, pharmaceutical industry, hospitals).
⊙ High-Performance Composites	Composites used for rackets, surfboards, helmets, pipes, windmill blades, aviation components.
⊙ Adhesives & Bonding	Adhesives (general purpose).
⊙ Consumer & Miscellaneous	Automotive parts (non-coating), Gardening tools and equipment, Printing inks.

1.12 Turtle

Turtles are among the oldest amphibian animals surveying on earth and are believed to have evolved approximately 220 million years ago. It functions as a protective armour plays a vital role in the survival and evolutionary success of the animal. Its delicate yet optimized architecture offers effective protection against external threats, while simultaneously serving as an anchoring platform for muscles and as primary mineral storage site within the body [Zhang, W et. al., 2012].

Turtles classified under the Order Testudines, are repetitious distinguished by a body enclosed in a bony shell, a characteristic that also includes tortoises. Although shells have evolved independently in a wide range of organisms from invertebrates to mammals the turtle shell exhibits a unique architectural design. It consists of an upper carapace and a lower plastron which are typically fused along the sides of a body to form a rigid skeletal box. This structure composed a bone and a catalyst precisely throughout the turtle's lifetime. As the shell is an integral component of the body, turtles cannot exceed it nor is it shed like the skin in some other reptiles.

Approximately 356 spices of tattoos inhabited terrestrial environments across all continents except Antarctica, as well as both fresh water and marine ecosystems. Tortoises, classified under the family Testudinidae, are exclusively land-dwelling and possess distinct anatomical characteristics that differentiate them from other turtles. However, the term "tortoise" has historically been applied to other terrestrial testudines including box turtles and wood turtles. In a similar manner the term terrapin was once used broadly to describe aquatic turtles but is now primarily associated with the edible diamondback terrapin found in the eastern United States [Turtle, 2020, Encyclopaedia Britannica]. Although turtles are widely distributed, the long evolutionary history has never been characterised by a large number of spices. Nevertheless, this relatively limited spices count does not imply a lack of diversity among turtles. Spices of Turtles are depicted in Figure1.10

Figure 1.10 Spices of Turtles (https://www.britannica.com/animal/turtle-reptile)

1.12.1 Structure and Properties of Turtle shell

Natural and artificial composites represent the two main categories of composite materials. Natural composites consist of plant stems such as timber and bamboo and animal products including bone and shell. These materials are referred to as in situ composites because the reinforcing elements form during the biological growth [Xu, Y et.al 1995].

In recent years, biological structural materials have become a significant focus of research due to the superior mechanical performance, which often surpasses that of comparable man-made materials. These outstanding mechanical properties arise from the unique organisation of material composition and structure. Biological materials consist of both organic and inorganic components integrated into complex architectures that are hierarchically arranged across the nanoscale, microscale and mesoscale full shop investigating biological materials and systems allows materials scientists and engineers to develop designs inspired by nature. This research area, known as biomimetics, represents one of the emerging frontiers in materials science. Numerous studies have been conducted to examine the structure and the mechanical behaviour of biological skeletons [Damiens. R et. al., 2012].

Turtle shells provide effective protection against predator attacks, offering valuable inspiration for the development of bioinspired synthetic materials. Despite being composed of a relatively weak constituents and exhibiting low levels of biomineralization, turtle shells display exceptional mechanical robustness due to their highly organised, layered architectures spanning from the nanoscale to the macroscale. The keratin-covered Bony exoskeleton displays several sectarian motifs such as altering rigid and compliant regions, layered architectures and functionally graded components all of which are decided to protect their appetite from predatory attacks and impact events. Mineral faces primarily calcium phosphate and calcium sulphate polycrystals are randomly distributed within a keratin matrix to form organic dash inorganic composites platelets. These plates serve as fundamental building blocks that are orderly and densely stacked in the radial direction forming a layered microsection configuration. The protective external structures of natural organisms provide significant insight into the development of synthetic materials with robust and multi-functional mechanical properties [Chen, M., 2017].Figure 1.11 show the skeleton structure of the turtle shell.

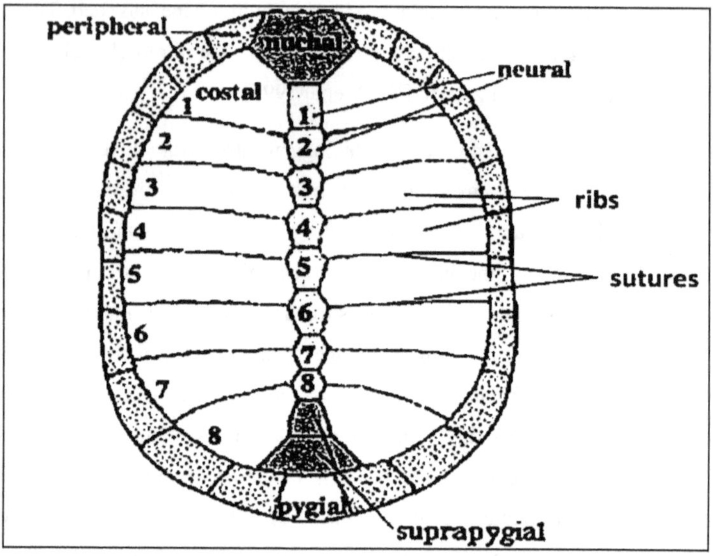

Figure1.11 Skeleton structure of the turtle shell [Achrai, B., 2017].

Analysis of inversion and self-righting processes in various turtle spices reveal that the geometry of the turtle shell is optimally designed to facilitate reversal movements each spices exhibit special shape adopted to the specific proportions of its neck and legs. The microstructure of the turtle shell is shown in Figure: 1.12. The natural shall structure represents a naturally multi-level system full sub in addition to enabling effective load bearing the evaluated shell design maximises the internal volume available for storing and protecting the animal's vital organs [Zhang, W et. al., 2012]. The mechanical properties of turtle shell are shown in Figure 1.13. Also Figure 1.14 illustrates the compression test of turtle shell. Table 1.3 shows that the microhardness of turtle shell.

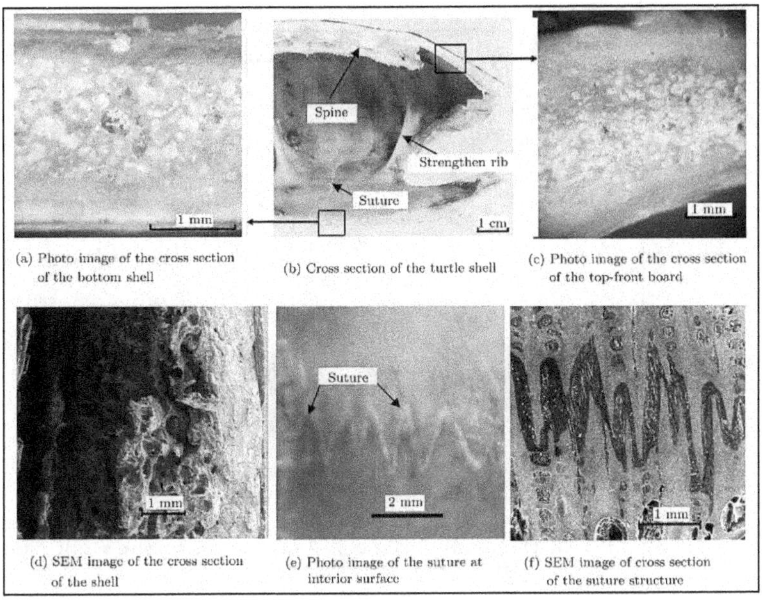

Figure 1.12A view of turtle shell microstructure [Zhang, W et. al., 2012]

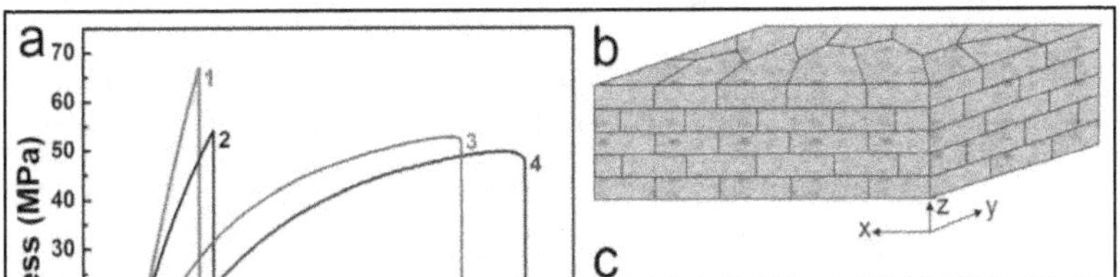

Sample	Modulus (GPa)	Strain Energy Density (MPa)	$Strain_M$ (%)	$Stress_M$ (MPa)
1	3.50	68.78	2.06	66.97
2	2.80	75.18	2.44	54.15
3	1.53	345.27	9.03	52.87
4	0.99	369.04	10.73	49.77

Figure 1.13 Mechanical Properties of turtle shell – Tensile. (a) Stress-Strain curve. (b) A model of turtle shell. (c) Young's modulus, strain energy density, tensile strain and strength of turtle shell [Chen, M., 2017].

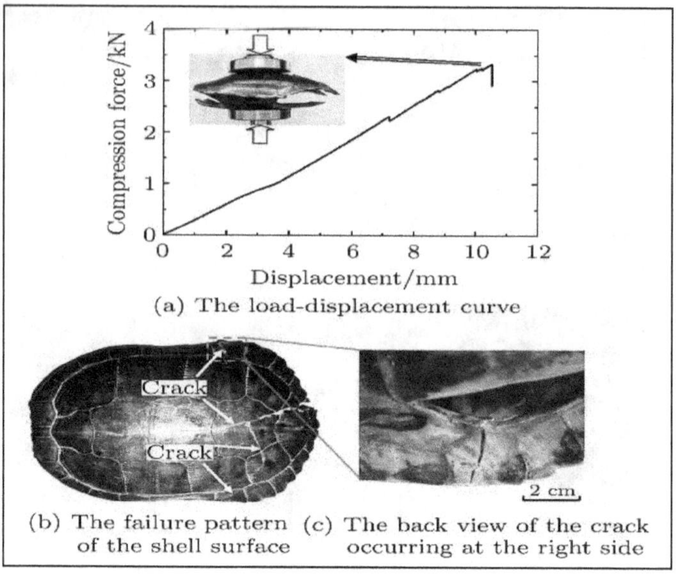

Figure 1.14 Compression test of the turtle shell [Zhang, W et. al., 2012]

Table 1.3 Microhardness of the turtle shell on different layers [Liang, Y. et.al., 2012]

Position	**Compact Layer**	**Spongy Layer**	**Inner layer**
Microhardness [HV]	51.38	32.3	48.34

1.12.2 Applications

The unique structure of the turtle shell - their natural composite of bone the skewed and acquaintance exterior - as led to various specialist applications in both its solid form and its processpowder form. The applications are listed below [Lo,C.F., et.al., 2006, Hainschwang, T., et.al., 2006, Li, L., et.al.,2012, Ovhal,D.A., et.al., 2020]

1. **Special applications of turtle shell powder.**

⊙ **Biomimetic Inspiration:** Engineers study the sutures joints to design the high- impact resistant materials and armor. The shell is incredibly efficient at dissipating energy from predators.

- **Artisan and Luxury goods:** Historically, tortoise shell was used for high end optical frames combs jewellery and furniture in lace because of its beautiful method appearance and ability to be polished to a high sheen.
- **Traditional Musical Instruments:** In various cultures have been used as a resonator for stringent instruments (like African kora or ancient iyres) due to the natural acoustic properties.
- **Dosimetry Applications:** It involve measuring and assessing radiation doses across medicine (radiotherapy, nuclear medicine), industry (sterilization, food irradiation, facility monitoring).

2. Special applications of turtle shell powder

- **Traditional Chinese Medicine (TCM):** Know as Gui Ban, the powder is used in traditional formulas. It is believed to nourish yin, support kidney health, and improve bone strength due to its high calcium and collagen content.
- **Paramedical Research:** Studies have explored turtle shell powder as a source of natural hydroxyapatite and bioactive peptides. It is sometimes research for its potential in treating osteoporosis or promoting bone tissue regeneration.
- **Cosmetics and Skin Care:** Due to high concentration of collagen and amino acids, the powder is occasionally incorporated to high end anti-ageing creams and masks marketed from its skin-firming properties.
- **Nutritional Supplements:** In some regions, it is processed into capsules as a natural calcium and trace mineral supplement.
- **Biomaterial and composite fillers:** Turtle shell powder rich in calcium based minerals can be explored as bio-derived filler in polymer composite to enhance stiffness, hardness and wear resistance.

1.13 Fibers

Fibres play a crucial role as the primary loan bearing component in polymer composites and shop a wide variety of fibres are available and are broadly classified into two categories natural fibers and synthetic fibers.

Natural fibers are the fibres derived from the natural resources, and it was for the classmates to three groups: animal – wool, silk, hair, Minerals – asbestos, Cellulose or ligncellulose – bast, leaf, reed/grass.

Fiber name	Plant	Fiber
Jute (Bevitori*et al.* 2010)		
Hemp (Hasan *et al.* 2020; Sanjay *et al.* 2016)		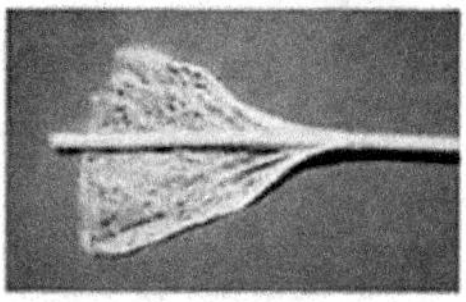
Flax (https://www.slideshare.net/Farhanullahbaig/flax-fiber, Hasan *et al.* 2020)	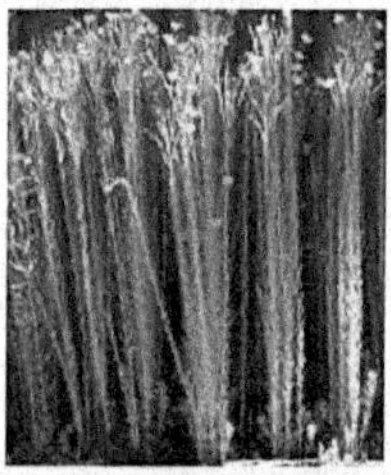	
Kenaf (Lam and Yatim, 2015)		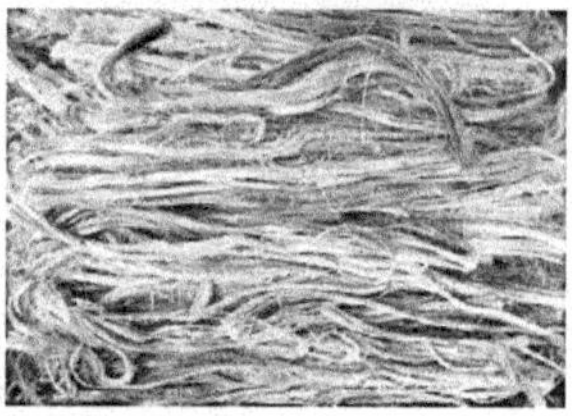
Sisal (Hasan *et al.* 2020; Sanjay *et al.* 2016)		
Rosella (en.wikipedia.org)		

Banana
(Sanjay *et al.* 2016, en.wikipedia.org)

Bamboo
(Glória et al, 2015)

Basalt
(Dhand*et al.* 2015; https://www.princelund.com/basalt-fiber.html)

Figure1.15 Source of Natural fibers a view. [Godwin Pithalis, NE., 2022].

Bast fibres such as jute, hemp, flax, kenaf, ramie, okra, rattan, and wisteria are extensively used in composite applications [Fan, M, et.al., 2016]. Leaf fibres including sisal, banana, abaca, and agave are also widely utilized [Fan, M, et.al., 2016]. Reed fibres such as bamboo, bagasse, communis, phragmites, and esparto have gained significant attention for composite fabrication [Petroudy, 2017]. The natural fibres commonly employed in fibre-reinforced polymer composites are illustrated in Figure1.15.

1.13.1 Basalt fiber and its applications

Basalt fiber is obtained from the volcanic arc originating from solidified lava and is categorised as a natural favour. It is widely recognised as a viable alternative to conventional class fibers. Basalt Fiber possess high Thermal stability, with melting temperatures ranging from 1500 to 2700 °C. Although the manufacturing procedure closely resembles that used for glass fibers, Basalt fiber Production consumes comparatively less energy. During fabrication, basalt rocks are heated to approximately 1500 degrees Celsius to produce molten basalt, which is subsequently transferred into fibers through a continuous spinning technique. The final fibers are supplied in Either continuous filament or chopped fiber form.

A variety of processing methods are used for the manufacture of basalt fibers, including blowing melt technology, the spinneret method, and the melt-spinning process. The conventional processing layout for producing continuous basalt fibers through the spinneret method is depicted in Figure 1.16. Figure 1.17 illustrates the chemical composition of basalt fibers, which are rich in silicon, aluminium, iron, magnesium, and calcium oxides (Fiore et al., 2015). The presence of these major chemical constituents significantly enhances the mechanical, thermal, and chemical properties of basalt fibers, as shown in Figure 1.18.

Applications:

Basalt fiber is widely used across various industries due to Its excellent mechanical, thermal and chemical properties. Key applications include:

- Construction and infrastructure
- Automotive and transportation
- Aerospace and defence
- Marine applications
- Electrical and electronics
- Thermal and fire protection
- Chemical and industrial uses
- Energy sector
- Sports and Leisure
- Geotechnical and Environmental engineering

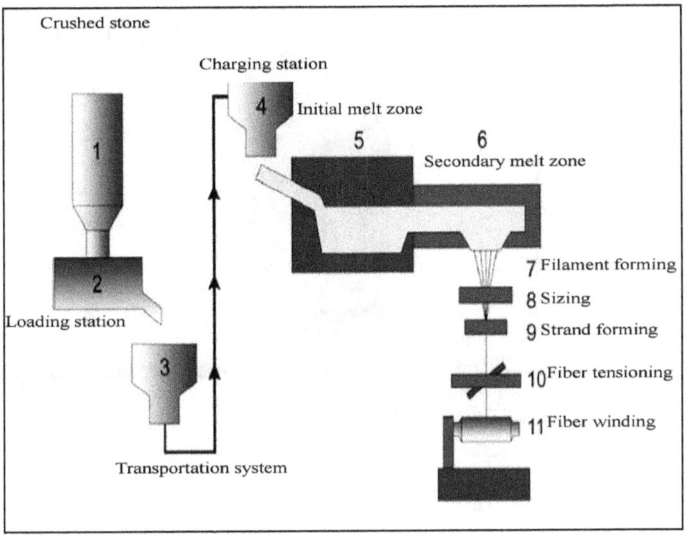

Figure1.16 Schematic layout of processing the basalt fiber (Fiore etal. 2015)

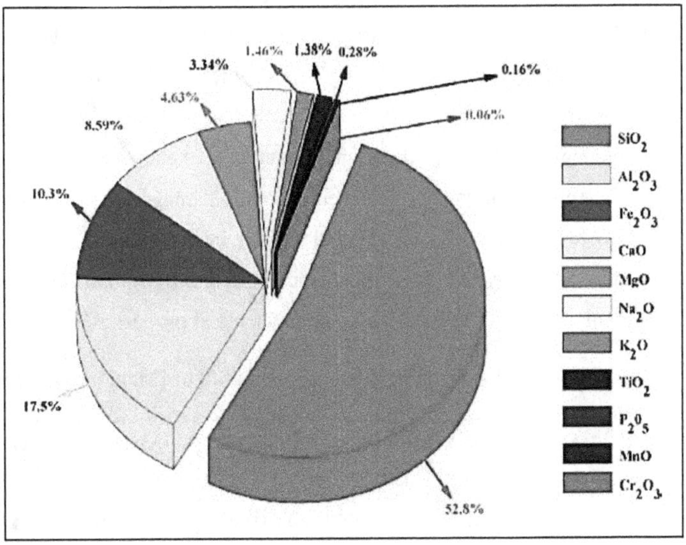

Figure 1.17 Basalt fiber – Chemical elements (Dhand et al. 2015)

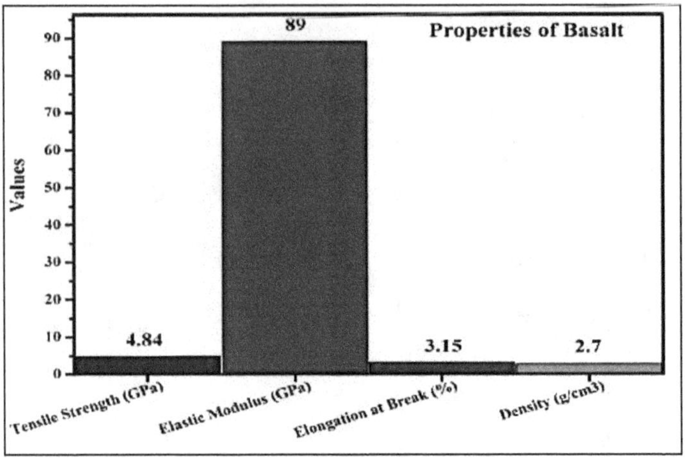

Figure1.18 Typical properties of Basalt fiber (Dhand et al. 2015)

1.13.2 Kenaf fiber and its applications

In recent years, the growing utilisation of a plant bastfibers to satisfy the increasing demand for natural fiber based composites has enhance the significance of kenaf, leading to its widespread cultivation across the globe. Kenaf fibers demonstrate a high cellulose content, which contributes to the biodegradability. An important advantage of kenaffiber is the absence of silica, thereby minimising abrasiveness and damage to processing equipment. However, studies have indicated that the incorporation of a ligocellulosic fillers may adversely affect the mechanical strength of composite materials [Aji, I.S., et.al., 2009].Table 1.4. Shows the mechanical properties of the kenaf fiber.

Table 1.4 Mechanical Properties of Kenaf fiber [Mahjoub, R. et.al., 2014].

S.No.	Properties	Measurements
1	Density (g/cm3)	1.4
2	Tensile Strength (MPa)	284-800
3	Tensile Modulus (MPa)	21-60
4	Elongation (%)	1.6

Kenaf fiber has been identified as an important fiber source for composite materials and various industrial applications. Its Renewability, biodegradablility and low environmental impact make it an important material for sustainable and green technologies.A notable advantage of its high fiber yield, with up to 40% of the stack providing usable fiber-approximately twice that of Jute, hemp or flax- making it a cost effective option. Recent developments have expanded kinaf applications to paper products construction materials absorbance and animal feed alongside a growing interest in its use as a reinforcement in composite materials [Anandjiwala, R.D., et., 2007].

Kenaf has been identified as a promising fibre material for the future due to its rapid growth rate, wide ecological adaptability and zero-waste potential[Juliana, A.H.et.al., 2012].The properties of kenaf fiber depend strongly on its microstructure, especially the cell wall composition. Cell wall polymers under their matrix control the physical and chemical behaviour of the fibre. Kenaf fibre changes size when moisture content varies because hydroxyl and other oxygen-containing groups in the cell wall attract water through hydrogen bonding. Hemicelluloses mainly cause moisture absorption, but cellulose, lignin and crystalline cellulose surfaces also contribute [Rowell, R.M.,et.al., 1999].The chemical composition of kenaf fiber is listed in table 1.5.

Table 1.5 Different stages of chemical composition of kenaf fibers [Khalil, H.A., et.al., 2010].

Chemical Composition	Kenaf (core + bast)	Kenaf core	Kenaf bast	Softwood	Hardwood
Extractive (%)	6.4	4.7	5.5	0.2–8.5	0.1–7.7
Holocellulose (%)	87.7	87.2	86.8	60–80	71–89
α-Cellulose (%)	53.8	49	55	30–60	31–64
Lignin (%)	21.2	19.2	14.7	21–37	14–34
Ash (%)	4	1.9	5.4	<1	<1

Microscopic examination of ground bark reveals the intricate cell wall structure of plant fibres, as illustrated in figure 1.19(a). and the schematic in figure 1.19(b). These structures, commonly referred to as macrofibrils, are characterized by a hollow central channel known as lumen, surrounded by four distinct layers: a single primary wall and a triple-layered secondary wall. Each of these layers is a composite material where cellulose is integrated into a lignin and hemicellulose matrix. Furthermore, As shown in figure1.19(c) The Microfibrils themselves consist of an alternating argument of crystalline and amorphous sections [Baillie, C.,et.al., 2004].

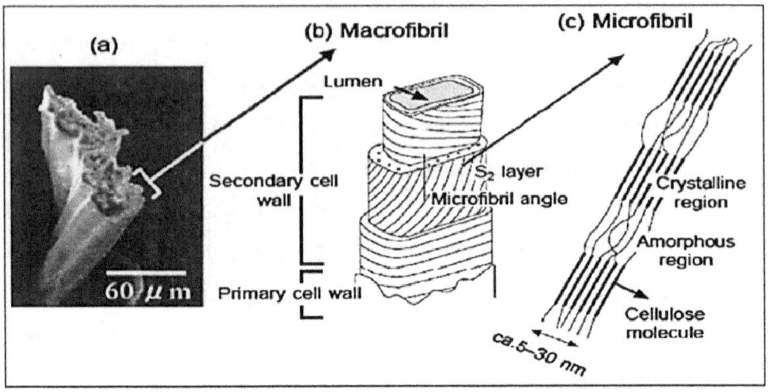

Figure1.19 Schematic representation of kenaf fiber (a) SEM image, (b) Macrofibril structure, (c) Microfibril structure[Baillie, C.,et.al., 2004]

Applications:

Kenaf fibre (Hibiscus cannabinus) is a fast growing, sustainable natural fibre with good mechanical strength, low density and high cellulose content. Because of these properties, it is widely used in many fields. Below is a list of potential applications for kenaf fibres and composite materials derived from them [Ramesh, M., 2016]

- Rope, twine, cordage, etc.,
- Engineered wood;
- Bags and fabrics, hamburger wrappers, fast food containers, wallpaper;
- Sails;
- Coarse cloth (similar to that made from jute);
- Energy (both electricity/heat and II generation biofuels);
- Animal bedding, poultry and cat litter;
- Feeds for its high protein and good digestibility (even 3 cuts/year);
- Insulated panels;
- Oil and liquid absorbent materials;
- Materials for blending with resins for bioplastic composites;

- Black liquor, a by-product of making paper from kenaf, is burned for fuel or chemical recovery in USA.
- Seed oil is edible and high in omega polyunsaturated fatty acids, well known compounds for reducing cholesterol and heart diseases.
- Automotive interior lining, wallboards, ceilings and furniture.

1.14 Outline of the thesis

The completed research findings are documented in the final thesis which is structured into six chapters. A brief overview of the content of each chapter follows.

Chapter 1 (Introduction): This chapter provides typically sets the stage for the entire study, providing background information of composite types, classification of polymer composites, fiber reinforced polymer composites, types of matrices, fillers and fiber, establishing the scope and significance of the work.

Chapter 2 (Literature Review): This section involves a critical review and synthesis of existing scholarly publications, previous research and relevant theories on fillers used in basalt and kenaf of FRP composites related to the thesis topic establishing the current state of knowledge and identifying the research gap addressed by the current study.

Chapter 3 (Materials, Methods and Experimental Techniques):This chapter deals the entire research approach of FRP composites, including the specific materials used, the experimental setup, the procedures and standards followed, mechanical and morphological characterization studies of FRP composites and the methods used for data collection and analysis.

Chapter 4 (Characterization of basalt / epoxy / TSP based FRP composites): This core chapter presents the raw and processed data obtained from the experiments of fabricated FRP composites. It often includes tables and figures and statistical outputs to symmetrically report the findings. It analyses the meaning of findings, discusses how they relate to the existing literature addressed whether the research objectives were met and explains the implications and theoretical contributions of the work.

Chapter 5 (Characterization of kenaf / epoxy / TSP based FRP composites): This core chapter presents the raw and processed data obtained from the experiments of fabricated FRP composites. It often includes tables and figures and statistical outputs to symmetrically report the findings. It analyses the meaning of findings, discusses how they relate to the existing literature addressed whether the research objectives were met and explains the implications and theoretical contributions of the work.

Chapter 6 (Conclusion and Future scope): The final chapter summarises the main outcomes of the research, states the final conclusions drawn from the study and suggests potential areas for future work or research that could build up on the current findings.

1.15 Summary of introduction

This chapter provides the fundamental background and motivation for composite materials. It systematically covers the types and functions of reinforcements, matrix materials and fillers before specifically detailing polymer composites, fiber reinforced polymer (FRP) composites some of their fabrication processes and their wide range of applications.

CHAPTER 2

LITERATURE REVIEW

2.1 INTRODUCTION

The objective of the literature survey is to study existing research related to the proposed research. It helps identify the fundamental theories, methodologies and technologies previously used in similar studies. This review supports the selection of appropriate design principles, analytical techniques and experimental approaches. Also, it provides a comparative understanding of different solutions implemented by earlier researchers. The survey assists in guiding the direction of investigation. Overall, it ensures that the proposed work is built on a strong theoretical foundation and contributes meaningfully to existing literature.

This chapter provides an overview of the relevant literatures related to various fillers used in FRP composites and fillers used in Basalt, Kenaf FRP composites. Also, the literature survey identifies the existing research gaps for the present work.

2.2 FILLERS USED IN POLYMER MATRIX COMPOSITES

Vinay et al., (2021) showed that adding nano-Al_2O_3 to basalt fiber–epoxy composites markedly influences their mechanical and wear behavior. Tensile strength generally declined with filler addition, though a slight recovery was noted at 0.6 wt%. Small amounts of nanofillers greatly improved flexural strength and interlaminar shear strength, with 0.4 wt% yielding a 31% rise in flexural strength and around 23% higher inter laminar shear strength (ILSS). Wear resistance increased steadily as nanofiller content rose, with 2.4 wt% reducing wear loss by nearly 40%. Excess filler, however, caused agglomeration and blow holes, weakening flowability and interfacial bonding. Impact strength and hardness improved up to 1.2 wt% filler loading, with hardness reaching its peak at 2.4 wt%. The best impact performance, around 36% higher than unfilled composites, also occurred at 1.2 wt%. SEM observations confirmed better mechanical and tribological behavior at lower filler contents. Overall, thermal stability remained largely unchanged regardless of nanofiller addition.

Pithalis et al., (2022) demonstrated the successful fabrication of basalt-fiber epoxy composites reinforced with varying amounts of silicon carbide using the hand-layup method. As the SiC content increases, the composite density and hardness rise, with the highest hardness of 92.4 HV observed at 10 wt% filler loading. Water absorption increases when more SiC is added, showing that the material takes up more moisture. Tensile strength improves up to 7.5 wt% SiC, reaching a maximum of 221 MPa, because the matrix is strengthened effectively. However, adding more SiC beyond this reduces the tensile strength. Flexural strength decreases as SiC content increases, with the highest value of 450 MPa observed in the composite without any filler. Fracture analysis shows features such as cleavage steps, delamination, fiber breakage, pull-out and cavities caused by filler debonding. Delamination is a main reason for failure under bending, leading to lower flexural strength. Overall, adding a controlled amount of SiC improves hardness and tensile properties, but too much filler reduces bending strength and structural stability.

Soundhar et al. (2019) developed hybrid composites made of sisal fiber and crab shell particles in an epoxy matrix using compression molding. The sisal fiber had a crystallinity of 63.76% and a crystallinity index of 0.43.It is showing mainly crystalline cellulose with some amorphous lignin. Crab shell particles had a hydrated crystalline structure with a major peak at 20° and an amorphous peak at 9.54° with a crystallinity of 63.95% and index of 0.44. Among the samples, the A3 hybrid composite showed the highest tensile strength of 46.70 MPa and flexural strength of 136.21 MPa. SEM images confirmed strong bonding between fibers and the matrix. It means the composite formed effectively. But adding more than 4 wt.% crab shell particles caused particle clustering and it reduces the mechanical performance of the composites.

Vivek et al. (2019) studied sugarcane bagasse ash (BGA-B) from three industries and found it could be used as recycled filler in epoxy composites. Its high silica content and good chemical, physical and thermal properties make it suitable for this purpose. BGA-B was added at 1, 3 and 5 wt.% to hybrid composites such as banana/flax, banana/kenaf, sisal/flax and sisal/kenaf. Adding BGA-B improved both mechanical and thermal properties. The highest flexural strengths were observed at 5 wt.% for sisal/flax (56.78 MPa) and banana/kenaf (34.2 MPa). Banana/flax and sisal/kenaf composites reached peak flexural strength at 3 wt.% BGA-B. Tensile strength showed a similar trend. Banana/kenaf and sisal/kenaf performed best at 5 wt.%, while banana/flax and sisal/flax were strongest at 3 wt.%. However, increasing BGA-B content reduced impact strength in sisal/kenaf and sisal/flax composites. Thermogravimetric analysis showed HBK composites were stable at 1–3 wt.% filler and HSK composites were stable at 5 wt.%. Overall, using low-cost bagasse ash improved thermal stability, reduced material cost and shows potential for marine, automotive and outdoor applications.

Ramesh et al. (2023) showed that waste granite and marble powders can be used as fillers in epoxy-based polymer composites. Using a hand lay-up method, adding these mineral fillers improved the physical and mechanical properties of the composites. The best results were achieved with 20% granite and 10% marble. At this ratio, the density reached 1.404 g/cm³, hardness was 87.65 BHN and compressive strength was 131.31 MPa. Hardness and compressive strength increased by 28.8% and 77%, respectively. The addition of granite and marble also improved wear resistance, making the composites more durable. These results show that low-cost industrial waste fillers can strengthen epoxy composites. Also make them suitable for wear-related applications.

Chandini et al. (2008) studied on tamarind seed. It is a byproduct of tamarind pulp processing. The decorticated seeds are used as a cheap sizing agent in jute and textile industries. But they have limited applications in other sectors. Research shows that tamarind seed powder has good gelling and adhesive properties. It can be useful in food and pharmaceutical products. Several studies and patents support these uses. However, its wider industrial potential has not yet been fully explored. Tamarind seed powder can be used in new and innovative products by better processing.

Olumuyiwa et al. (2012) studied coconut-shell-reinforced polyethylene composites to discover their use in engineering applications. The composites were made by mixing low-density polyethylene with 5–25% coconut shell particles. Results showed that hardness increased as more particles were added. However, higher filler content reduced tensile strength, modulus, impact energy and ductility. SEM images showed weak bonding between the coconut shell particles and the polyethylene matrix. Despite these limitations, the study shows that coconut shell, an agricultural waste, can be used as a low-cost filler in composites.

Amal et al. (2015) found that chicken eggshell (ES) has properties similar to commercial calcium carbonate. This makes it suitable as a filler in polymer composites. ES has good particle properties, thermal stability and composition. Its abundance and low cost make it attractive for large-scale applications. As it is lightweight, the ES is suited for low load bearing components. Overall, ES is a promising, eco-friendly filler and a sustainable resource for bulk composite production.

Saba et al. (2017) reported that the addition of nano oil palm empty fruit bunch (OPEFB) filler affects the density, crystal structure and thermal expansion of epoxy composites. As the filler content increased, the density of the composite also increased, indicating improved stiffness and hardness due to the rigid nano OPEFB particles. XRD analysis showed a reduction in peak intensity and crystallinity, which helped reduce the brittle nature of pure epoxy. This change was most effective at 3% filler content, where the structural modification was optimal. All the nanocomposites showed lower thermal expansion compared to neat epoxy, with the 3% filled sample giving the best result. Overall, the study concluded that 3% nano OPEFB provides the suitable balance between structural improvement and thermal stability.

Hom et al. (2018) showed that date palm fibres obtained from agricultural waste can be effectively used to reinforce polycaprolactone (PCL) biocomposites. The addition of these fibres improved both tensile strength and impact resistance. The study also found that higher extrusion screw speeds, such as 400 rpm, resulted in shorter fibre lengths compared to lower speeds like 100 rpm. The results highlight that date palm waste fibres can be used to produce lightweight, low-cost and sustainable biocomposites. Their use helps reduce environmental impact and supports better resource utilization. These materials offer good mechanical performance while lowering manufacturing costs, making them suitable for industrial applications.

Okele et al. (2018) studied the use of hybrid filler made from carbon black and cashew nut shell powder (CNSP) to reinforce natural rubber. The aim was to reduce material cost by partially replacing carbon black with agro-waste particles without losing mechanical strength. The results showed that CNSP performed well as reinforcing filler when combined with carbon black. Smaller CNSP particle sizes further improved reinforcement efficiency. The mixture containing 5 g of carbon black and 25 g of CNSP produced the highest tensile strength. This study confirms that CNSP is a low-cost, renewable alternative to carbon black and that hybrid fillers can provide a sustainable solution for rubber reinforcement.

Shravanabelagola et al. (2020) investigated the effect of alkali treatment on Limonia acidissima shell powder (LASP) for use as polymer filler. Alkali treatment reduced the hydrophilic nature of the powder by removing hemicellulose, lignin, and surface waxes, as confirmed by chemical and FTIR analysis. Microstructural observations showed a cleaner and more uniform surface after treatment. Proximate and elemental analyses revealed increased fixed carbon and total carbon content, improving its reinforcing potential. XRD results indicated higher crystallinity and larger crystal size, which enhanced moisture resistance. Thermal analysis using TGA and DTA showed slower cellulose degradation, indicating better thermal stability. Overall, alkali-treated LASP exhibited improved structural and thermal properties, making it suitable for polymer composite applications.

Hemabarathy et al. (2014) prepared porous three-dimensional scaffolds using alginate mixed with nano cockle shell powder. They observed that changing the alginate content directly affected the pore size and structure of the scaffold, which in turn influenced its shape and performance. The scaffold containing 40% alginate and 60% nano cockle shell powder showed the best balance of strength, stability, and structural integrity. This composition also supported better cell growth, indicating good biocompatibility. Because of these properties, the scaffold is suitable for tissue regeneration and shows strong potential as a bone graft material. However, further studies on cell attachment and in vivo behavior are needed to confirm its long-term performance.

Nitin et al. (2013) investigated the use of walnut shell particles as a natural reinforcement in polymer composites. The addition of walnut shell particles reduced the overall density of the composites, making them lighter and more suitable for lightweight applications. A sharp decrease in density was observed when the filler content exceeded 30 wt.%. Even with this reduction in weight, the composites maintained acceptable mechanical strength, with tensile strength values between 33 and 41 MPa. The study shows that walnut shell particles are effective and low-cost natural filler. They help achieve weight reduction while still retaining sufficient strength, making them a sustainable option for lightweight structural materials.

Fiore et al. (2014) found that epoxy composites with Arundo donax fillers (ADFs) had higher stiffness, with tensile modulus increasing by up to 40% compared to neat resin. Flexural stiffness improved only slightly and tensile and flexural strengths generally decreased due to weak bonding between the hydrophilic filler and hydrophobic epoxy. Filler size and content affected strength mainly through void formation. Dynamic mechanical analysis showed no major change in the plastic region, but storage and loss moduli increased in the rubbery region with larger filler size and higher loading. Damping behavior did not change much, indicating little restriction on polymer chain movement. SEM images showed poor filler–matrix adhesion and variations in morphology depending on filler size. Overall, Arundo donax fillers can improve stiffness in semi-structural epoxy composites while supporting sustainable use of this biomass.

Saba et al. (2017) showed that cellulose nanofibres (CNFs) from Northern Bleached Softwood Kraft (NBSK) pulp can be added to epoxy to improve its properties. The CNFs were nanosized and made of pure carbon and oxygen. Mechanical testing showed that the addition of CNFs increased the tensile, flexural, and impact strength of the composites. The best performance was obtained at 0.75% CNFs, where the particles were evenly dispersed without noticeable agglomeration. SEM and TEM images confirmed the presence of effective toughening mechanisms at this filler level. The results indicate that even a small amount of CNFs can be used instead of costly or harmful nanofillers. This offers a simple and sustainable approach to enhancing the properties of epoxy composites.

Alsaadi et al. (2018) studied polyester composites reinforced with microscale pistachio shell powder. The mechanical properties changed depending on the amount of filler. SEM images showed good particle distribution at 5 and 10 wt%, but higher amounts caused clumping. Tensile strength increased up to 10 wt% filler. Flexural strength peaked at 5 wt% but dropped at higher amounts, while flexural modulus continued to rise up to 25 wt%. Impact strength and energy absorption were highest at 5 wt% before decreasing. Overall, pistachio shell powder can be used to make eco-friendly and low-cost particulate composites.

Ambali et al. (2019) found that higher curing temperatures improved flexural, tensile and impact properties of cow-horn-reinforced polymer composites. Samples with 100 μm particles cured at 80°C had higher tensile and impact strength. Composites with 150 μm particles were best for applications needing high impact resistance. These results show that cow horn can be effective filler when cured at higher temperatures. The study suggests further research on epoxy composites with cow horn, including how alkali treatment affects bonding between particles and the matrix.

Hitesh et al. (2020) studied the wear performance of epoxy composites reinforced with citrus limetta peel (CLP) fillers under different loads and sliding speeds. Adding CLP particles improved wear resistance. However, wear loss and friction increased with larger particle sizes and higher loads. Among the samples, CP1 (100–250 μm) had the lowest friction, showing better wear resistance than CP3 (650–800 μm). SEM images showed microcracking, pit formation and debris as the main wear mechanisms.

Overall, adding CLP fillers improved the wear properties of epoxy. Using 15% food-waste filler also provides a sustainable way to reduce plastic use.

Mohammed Nadedm et al. (2021) examined hybrid epoxy composites reinforced with coir fiber and TiC nanoparticles. These composites showed improved physical, mechanical and thermal properties. Adding TiC increased density but also raised void content, which can reduce performance. Wettability tests showed lower contact angles in coir/basalt fiber/synthetic epoxy (CBSE) and coir/basalt fiber/TiC (CBABE33) laminates. It is indicating stronger surface interactions. Tensile strength increased in all hybrids due to good dispersion of coir and TiC particles. The CBSE laminate had the highest impact resistance, because of strong bonding between matrix and fillers. Hardness improved in CBASE composites because of hard reinforcements like basalt fiber and TiC. Thermal tests showed CBABE33 and CBASE retained more mass and had better thermal stability. Overall, these hybrid composites, especially basalt fiber-based laminates are promising, cost-effective and high-performance options for automotive and aerospace applications.

2.3 FILLERS USED IN BASALT FRP COMPOSITES

Lopresto et al. (2011) explored the possibility of using basalt as a filler instead of glass in epoxy composites. Mechanical test results showed that basalt-reinforced laminates can be a good alternative to E-glass laminates. Basalt fibres showed a much higher Young's modulus, about 35–42% higher than glass fibres. Basalt laminates also exhibited better compressive and flexural properties. However, glass fibre composites showed higher tensile strength. The stiffness of both composite systems remained consistent across different tests, suggesting that impact variations were mainly due to fibre volume differences. Distinct bending failure modes were noted, with glass fibres failing in compression and basalt fibres failing in tension. Short-beam strength values for both materials were nearly identical when normalized by density. This similarity indicates that the fibre–matrix bonding of basalt–epoxy composites is comparable to that of traditional E-glass–epoxy systems. Although the influence of areal weight differences is unclear, density-normalized data supported the superior performance of basalt fibres.

Eslami Farsani et al., (2014) reported the effect of adding micro glass powder into basalt fiber reinforced epoxy composites. This report introduces a new basalt-fiber-based FML (BFML) and evaluates its Charpy impact performance. The addition of MGP enhances the energy absorption capacity at both 25 °C and 75 °C, with a more pronounced improvement at higher temperatures and also increases flatwise impact energy at room temperature. Temperature variation has a smaller influence on EAC in $[0]_4$ composites than in $[45]_4$ laminates, where the former shows only an 8.5% rise while the latter exhibits a much larger 169.4% increase for the same 50 °C change. FML composites demonstrate significantly greater damage tolerance than plain BFRE in both edgewise and flatwise impact modes, especially when suitable metals are used in their fabrication. Moreover, FML specimens maintain structural integrity better than BFRE, making them suitable for primary structural applications. Overall, basalt-fiber composites clearly outperform glass-fiber composites under impact loading. It indicate that the basalt fibers are a promising alternative for impact-resistant composite structures.

Ary Subagia, I.D.G et al., (2014) examined basalt–carbon/epoxy hybrid composites. It is showed that the stacking sequence plays a critical role in determining their flexural performance. Hybrid laminates were fabricated using vacuum-assisted resin transfer molding with four basalt and six carbon fabric layers. Mechanical testing was carried out according to ASTM standards and fracture behavior was examined using microscopy. Compressive failure was identified as the dominant failure mode. Higher flexural strength and modulus were obtained when carbon layers were placed on the compressive outer side. Among all stacking sequences, the H1 (C3B4C3) configuration showed the best performance. It provided a 73% improvement compared to basalt fiber composites and was only about 14% lower than carbon fiber composites. Carbon layers helped distribute fatigue stresses and improved flexural behavior, while basalt layers on both tension and compression sides increased ductility. Laminates with carbon on the compressive side showed tensile modulus values above 40 MPa, whereas those with basalt on that side remained below this value. All stacking sequences showed a positive hybrid effect, offering a good balance of strength, ductility and cost. This study shows that mechanical properties can be modified by adjusting the stacking sequence.

Sampath et al. (2011) fabricated basalt fiber reinforced epoxy composites using the hand lay-up method with moringa ash and bagasse ash as fillers. Among the samples, Composite C containing moringa ash showed the highest open-hole tensile strength. It was 17% higher than Composite B with bagasse ash and 47% higher than Composite A without fillers. Composite C also showed the highest impact strength, mainly due to the silica-rich nature of moringa ash. Both filler-based composites exhibited improved resistance to acidic corrosion. Composite C showed lower water absorption, indicating stronger fiber–matrix bonding. SEM images shows fracture features, internal cracks and interfacial characteristics that supported the mechanical test results. Overall, the addition of ash fillers improved the strength, durability and environmental resistance of basalt fiber composites.

Nahid Jamali et al. (2019) studied the effect of graphene oxide (GO) and silane-treated graphene oxide (SGO) on basalt fiber reinforced epoxy composites. The addition of SGO considerably improved the mechanical properties of the composites. The composite with 0.5 wt.% SGO showed increases of 46% in tensile modulus, 54% in flexural modulus and 66% in compressive modulus compared to the neat composite. The highest strength improvement was observed at 0.4 wt.% SGO, with tensile, flexural and compressive strengths increasing by 18%, 59% and 61%, respectively. The experimental compressive results were in good agreement with theoretical predictions. Silane functionalization further enhanced the performance of GO-filled composites. The composite containing 0.4 wt.% GO showed strength improvements of about 24–26% and modulus increases of up to 39%. Overall, the addition of GO and SGO improved interfacial bonding and fracture behavior, resulting in mechanically improved basalt fiber epoxy nanocomposites.

Abd El-Baky et al. (2020) investigated the mechanical behavior of flax fiber composites hybridized with glass and basalt fibers. Among all laminates, the G6 configuration showed the highest tensile, flexural and impact strengths.

In contrast, the F6 laminate exhibited the highest strain to failure. Hybrid flax–glass–basalt composites showed clearly better performance than pure flax composites. The addition of glass or basalt fibers improved the tensile, flexural and impact strengths of flax composites and the improvement increased with higher synthetic fiber content. Pure flax laminates showed higher void content, while hybrid laminates exhibited lower voids due to better compatibility with the epoxy matrix. The stacking sequence strongly influenced performance. Laminates with glass skins and flax cores performed better in flexural and impact loading, while flax skins with glass cores showed better tensile properties. Overall, hybridization provided a good balance of strength, weight and cost, making these composites suitable for medium-load applications.

Saurabh Khandelwal et al. (2020) reviewed polymer and cement composites reinforced with basalt fibers. The study discussed methods to improve fiber–matrix bonding, such as surface treatment of fibers and the use of micro- and nanofillers. Compared to many natural and synthetic fibers, basalt fibers offer high strength, good chemical stability, strong thermal resistance and lower production cost. In polymer composites, basalt fibers improve flexural strength, impact resistance and fracture toughness. In cement-based materials, they enhance early strength, toughness and durability, although they can reduce workability due to higher water demand. Chopped basalt fibers also improve flexural behavior, but weak interfacial bonding limits their full effectiveness in cement composites. Various surface modification techniques have been explored to improve fiber wetting and surface roughness. Micro- and nanofillers further strengthen composites by improving interfacial interaction in polymers and densifying the cement matrix through secondary hydration reactions. Overall, basalt fiber is identified as a promising reinforcement for advanced composite applications.

Nick Wolter et al. (2021) compared different manufacturing techniques for basalt fiber reinforced polybenzoxazine composites. The methods included vacuum infusion, hand lay-up, dynamic fluid compression molding and autoclave curing. Composites produced using vacuum infusion showed mechanical properties equal to or better than those made using high-pressure autoclave and compression molding. Vacuum-infused composites exhibited about 10% higher tensile strength and 24% higher strain at failure compared to autoclave-cured samples. These results were achieved despite the low-pressure nature of the vacuum infusion process. The study showed that low porosity and high mechanical performance can be obtained without high-pressure curing. A strong relationship was observed between processing method, composite structure and mechanical behavior. Overall, vacuum infusion was identified as a simple, cost-effective method for producing high-quality basalt fiber composites.

Vinay et al. (2020) studied the effect of nano-Al_2O_3 particles on basalt fiber reinforced epoxy composites. The results showed that tensile strength generally decreased with filler addition, although a small improvement was observed at 0.6 wt.% filler content. Low amounts of nanofillers significantly improved flexural strength and interlaminar shear strength. A filler content of 0.4 wt.% increased flexural strength by 31% and ILSS by 23% compared to unfilled composites. Wear resistance improved continuously with filler addition and 2.4 wt.% nanofiller reduced wear loss by nearly 40%. However, filler contents beyond the optimal limit caused particle agglomeration, blow holes, poor resin flow and weak interfacial bonding. Impact strength and hardness improved up to 1.2 wt.% filler, with maximum hardness recorded at 2.4 wt.% and a 36% increase in impact strength at 1.2 wt.%. Thermal stability was only slightly affected by nanofiller addition. SEM analysis confirmed that lower filler contents improved mechanical and wear performance, while excessive filler led to defects and interface degradation.

Suresh Kumar et al. (2020) examined the effect of temperature on the low-velocity impact behavior and post-impact flexural strength of hemp/epoxy, basalt/epoxy and hemp–basalt hybrid composites. Hemp/epoxy composites absorbed the highest impact energy, while basalt/epoxy absorbed the least. Hybrid composites showed intermediate behavior. Impact energy absorption increased with temperature for all composites. At 50 °C, hemp and hybrid composites absorbed slightly more energy than basalt composites. Basalt/epoxy consistently showed the lowest absorbed energy and the highest toughness. Hybridization with basalt improved the impact resistance of hemp composites by reducing deformation and absorbed energy. Hemp/epoxy showed the lowest performance index and the highest absorbed energy and moment parameter. Temperature had little additional effect due to matrix softening. Basalt/epoxy exhibited the highest performance index and lowest absorbed energy at both room and elevated temperatures. Post-impact flexural tests showed that residual strength decreased with temperature, but hemp and hybrid composites retained most of their strength at 50 °C.

Fiore et al. (2017) studied the aging behaviour of jute laminates and jute–basalt hybrid composites under cyclic hygrothermal and UV exposure for 84 days. The hybrid laminates showed better flexural strength, impact energy and overall durability than pure jute composites. Among the hybrids, the sandwich configuration performed better than the intercalated structure. This was because the basalt outer layers protected the inner jute fibers from environmental degradation. The flexural and dynamic mechanical properties of sandwich hybrids were less affected by aging, indicating improved environmental stability. Hybridization reduced the limitations of jute fibers in outdoor conditions. Increasing the thickness of basalt skins further delayed degradation and minimized the loss in impact strength. But jute-only laminates showed the lowest mechanical performance throughout the aging period. Overall, sandwich-type jute–basalt hybrids were found to be the most suitable for long-term outdoor applications.

Leea et al. (2010) investigated the effect of carbon nanotube surface modification on basalt fabric reinforced epoxy composites. The study compared unmodified, oxidized and silane-treated CNTs. Composites containing silanized CNTs showed the best tensile performance. Tensile strength and Young's modulus increased by 34% and 60%, respectively, compared to unmodified CNT composites. These composites also showed improved thermal stability, with higher storage modulus and glass transition temperature. The improvements were attributed to restricted polymer chain movement and strong cross-linking within the matrix. Good dispersion of silane-treated CNTs and strong interfacial bonding further contributed to the enhanced performance. Overall, silane functionalization was identified as the most effective method for improving the mechanical and thermal properties of the composites.

Vikas Sharma et al. (2020) fabricated glass–epoxy and basalt–epoxy composites. He also studied their physical, mechanical and abrasive wear behavior. Fiber reinforcement significantly improved composite performance by enhancing bonding with the epoxy matrix. Both glass and basalt fiber composites showed lower abrasive wear rates with increasing fiber content. For glass fiber composites, the wear trend followed EGF-40 > EGF-30 > EGF-20 > EGF-10, while for basalt fiber composites it followed EBF-40 > EBF-30 > EBF-10 > EBF-20. Taguchi analysis identified abrasive size, sliding distance, applied load and fiber content as the most influential factors affecting wear. SEM observations revealed three-body abrasive wear mechanisms, including ploughing, fiber breakage and fabric damage. Overall, basalt fiber composites showed lower wear rates and performed better than glass fiber composites.

Seghini et al. (2019) investigated the interfacial adhesion of basalt fibers with different surface treatments in an epoxy matrix. The fibers were commercially sized, thermally treated, or coated using plasma polymerization with tetravinylsilane. Interfacial shear strength was measured using single-fiber fragmentation tests. Plasma-treated fibers showed the strongest fiber–matrix bonding. These fibers exhibited reduced debonding, shorter fragment lengths and higher interfacial shear strength values compared to the other treatments. The improved adhesion was mainly due to increased surface roughness and better mechanical interlocking with the epoxy matrix. Overall, plasma polymerization was found to be the effective method for enhancing interfacial bonding in basalt fiber epoxy composites.

Prasad et al. (2019) studied the mechanical behavior of jute–basalt hybrid epoxy composites with different stacking sequences. Hybrid laminates C4 (JBJJBJ) and C5 (JBBBBJ) showed lower void content compared to the pure jute composite C1 (JJJJJJ). In contrast, laminates C2 (BJJJJB) and C3 (JJBBJJ) exhibited higher void content. Moisture absorption was below 5% for all composites except C4. Tensile strength did not show a clear trend among the stacking sequences. However, composite C5 showed about a 20% increase in tensile strength compared to C1. Flexural strength improved with changes in stacking sequence. Impact strength varied across the hybrids and did not follow a clear pattern, with generally lower values than the pure jute composite. Overall, the hybrid laminates showed better mechanical performance than the all-jute composite. The results indicate that jute–basalt hybrid composites can serve as alternatives to glass fiber composites for medium-strength applications. Further studies are required to identify the most effective stacking sequence.

Bashar et al., (2013) proved that the nanoparticle modification can significantly influence the fracture behavior of brittle epoxy polymers. Nanoclay additives caused crack deflection and branching but did not produce major toughness improvements, whereas block-copolymer additives formed micelle structures that promoted cavitation and yielded substantial toughness gains. The results indicate that matrix ductility is the key factor governing delamination crack initiation when fibre-matrix bonding is strong and fibre constraints are minimal. Although nanoclay-modified composites exhibited fracture energies exceeding those of the bulk epoxy, block-copolymer systems did not fully translate their bulk toughness into composites due to fibre restrictions on plastic zone growth. Delamination toughness increased further through fibre-related mechanisms such as pull-out and bridging and was higher at reduced fibre volume fractions because of greater matrix deformation. Overall, the study confirms that beyond a certain threshold (~700 J/m^2), increases in resin toughness do not proportionally enhance interlaminar fracture toughness in fibre composites with similar fibre volume levels.

Danuta et al., (2019) demonstrated the incorporation of basalt powder into epoxy composites significantly enhances their thermal and thermo-mechanical behavior. The eco-friendly filler improves the resistance of the epoxy matrix to high temperatures and fire exposure. This leads to better thermal stability of the composite. Higher activation energy values and increased total heat release confirm improved resistance to thermal degradation. The close agreement between the Ozawa and Kissinger models supports the reliability of the thermal analysis. DTG results show that basalt powder slows down the degradation rate of the epoxy resin. The modified composites also show higher stiffness and hardness than the unfilled epoxy. Increases in storage modulus and crosslink density indicate stronger molecular interactions within the matrix. As the basalt content increases, the thermomechanical performance of the composites steadily improves. Overall, basalt powder acts as an effective and environmentally friendly modifier. It enhances both thermal stability and mechanical strength of epoxy composites.

Mohammad et al. (2013) studied the effect of nanoparticles on the fracture behavior of brittle epoxy polymers. Nanoclay particles slowed crack growth by causing crack deflection and branching. However, they did not significantly improve fracture toughness. In contrast, the addition of acrylic tri-block copolymers formed micelle structures that promoted cavitation and matrix deformation. This mechanism resulted in a clear improvement in toughness. The findings also suggest a qualitative link between fracture energy in bulk epoxies and their behavior in fibre-reinforced composites. Matrix ductility emerges as the primary factor governing delamination initiation, provided the crack-tip plastic zone is not constrained by adjacent fibres and the fibre–matrix bond remains strong. Nanoclay-filled epoxy systems even exhibited higher steady-state crack growth resistance than the neat polymer. Toughness was further elevated by fibre-related mechanisms such as pull-out, bridging and breakage. However, block-copolymer-modified systems did not fully transfer their bulk toughness improvements due to fibre restrictions on plastic zone development. Lower fibre volume fractions were found to enhance interlaminar toughness by allowing greater matrix deformation. Consistent with previous studies, the results confirm that beyond about 700 J/m^2 of resin toughness, further gains do not effectively translate to higher composite delamination resistance when fibre content remains constant.

Mohammed et al., (2021) developed hybrid bio-synthetic epoxy composites reinforced with coir fibres and TiC nanoparticles and their physical, mechanical and thermal behaviour was thoroughly evaluated. The incorporation of these reinforcements led to notable improvements in the overall performance of both bio-based and synthetic epoxy matrices. Experimental results showed that while TiC addition increased the actual density of the composites, it also raised the deviation from theoretical density, indicating increased void content, which is undesirable. Contact angle measurements showed that the CBSE and CBABE33 laminates had lower values. This indicates higher wettability, which may lead to reduced strength. Tensile strength increased in all reinforced samples due to the uniform dispersion of coir micro-particles and TiC nanoparticles in the epoxy matrix. The CBSE laminate showed the highest impact strength because of strong bonding between the fillers and the epoxy resin. Shore hardness also increased, with the highest value observed for the CBASE composite. This improvement is due to the presence of hard and brittle reinforcements such as TiC and basalt fibers. Thermal analysis revealed that the CBABE33 and CBASE laminates had better thermal stability. This was supported by higher residual mass and the ability of TiC particles to restrict molecular motion.

Velumayil et al. (2022) investigated the mechanical behavior of neat and hybrid laminates made of basalt, Kevlar and epoxy using the hand layup method. The laminates showed fiber volume fractions between 32% and 40%. It is typical for this process. Hybrid laminates showed mechanical properties between pure basalt and pure Kevlar composites. This trend was observed for tensile, flexural, impact strength, modulus and hardness. However, some hybrid configurations showed lower strength than the neat laminates, indicating the need for further investigation. The study also confirmed that fiber volume fraction strongly influences composite performance. Overall, the work highlights the potential of basalt–Kevlar hybrid composites and the importance of optimizing laminate design for improved performance.

2.4 FILLERS USED IN KENAF FRP COMPOSITES

Yousif et al. (2022) reported that the flexural performance of kenaf fibre reinforced epoxy composites depends on the fibre surface condition. NaOH treatment significantly improved fibre–matrix bonding and resulted in higher flexural strength. Treated fibres showed a 36% improvement and untreated fibres showed only a 20% increase. In untreated composites, failure mainly occurred due to fibre–matrix debonding, tearing, pull-out and fibre detachment. This behavior indicates weak interfacial adhesion. Also the treated composites failed mostly by fibre breakage, showing that the fibres effectively carried the applied load. The 6% NaOH treatment also split fibre bundles into finer fibres. This improved resin penetration and mechanical interlocking. Reduced porosity in treated composites further enhanced flexural performance. Overall, kenaf fibres showed excellent reinforcing ability under bending loads and demonstrated strong potential as a sustainable alternative to glass fibres.

Ramesh et al. (2016) discussed the growing demand for sustainable and renewable materials and highlighted kenaf fibre as a suitable option for engineering applications. Kenaf composites are used in textiles, automotive, construction and consumer products. The studies show that kenaf fibre reinforced composites have good mechanical and thermal properties. However, water retting using chloride-rich water can damage fibre structure and reduce strength. The source of retting water has little effect on tensile properties. Environmental exposure significantly reduces fracture toughness, depending on fibre orientation, void content and permeability. Excessive alkaline treatment, especially with 8% NaOH, weakens fibres and makes them brittle. Tensile modulus changes only slightly after treatment, but strain values vary widely. Aging environments reduce fibre strength, although treated fibres perform better than untreated ones. Wear behaviour depends on fibre orientation, load and sliding speed, with failure dominated by micro-cracking, debonding and matrix deformation. Heat deflection tests showed that kenaf composites have better thermal resistance than neat resin. DMA results confirmed higher storage modulus and improved fibre–matrix adhesion in treated composites, especially those treated with silane. Sound insulation results followed classical resonance theory. DSC studies revealed that moisture increases plasticity and reduces strength, while fibre length strongly affects crystallization behavior. SEM analysis confirmed fibre pull-out and fibre breakage as the main failure modes.

Hamdan et al. (2016) demonstrated that woven kenaf composites perform much better than randomly chopped kenaf composites. Improvements of up to 199% in flexural strength and 178% in flexural modulus were reported. Tensile strength and modulus also increased by 177% and 84%, respectively. Fibre orientation has important role in determining performance. Type A woven layers with higher yarn density, showed better mechanical properties than Type B layers with lower yarn density. Increasing yarn density improved load-bearing capacity, as seen in the superior performance of the A/A/A orientation compared to the B/B/B configuration. Failure analysis revealed fibre–matrix debonding, fibre pull-out and fibre detachment as the main fracture mechanisms. Type A woven composites also outperformed random kenaf composites, highlighting the importance of controlled fibre arrangement. Although woven kenaf composites still fall short of glass fibre composites, optimizing yarn orientation offers a promising route for improving their structural performance.

Sathish et al. (2017) studied the mechanical and thermal behaviour of banana–kenaf hybrid composites. It is confirmed that these materials are biodegradable and cost-effective compared to conventional composites. Among all samples, the composite with the highest kenaf content (Sample 3, 60%) showed the best performance. It recorded the highest tensile modulus and flexural strength and exhibited a clear linear relationship between load, stress and displacement. This sample also showed the highest double-shear strength and better energy absorption than the other composites. Due to its superior properties, Sample 3 was further tested under compression. The result showed the strength increased with laminate thickness. Thermal expansion results indicated more stable thermal behaviour. The TGA confirmed improved flame resistance. Overall, increasing kenaf content enhanced both the mechanical and thermal properties of the hybrid composites.

Siti Nor et al. (2018) developed woven kenaf–polyester composites and used the Taguchi method to optimise fibre orientation. It has improved mechanical performance. Because natural fibres show large property variations, optimisation through experimental design was essential. The Taguchi approach identified the best fibre angle combinations using fewer experiments. Tensile strength and Young's modulus were analysed using Minitab software. Experimental validation confirmed that the optimised fibre orientations produced higher mechanical strength. The study showed that fibre orientation is a important factor influencing composite behaviour. The each layer contributing to changes in strength and stiffness. Overall, the Taguchi method proved effective for improving the mechanical performance of woven kenaf composites.

Napisah et al. (2018) investigated the effect of carbon nanotube content and surface treatment on the mechanical properties of kenaf/epoxy composites. Composites containing silane-treated CNTs showed the best performance among all samples. Increasing silane-treated CNT content led to continuous improvements in tensile, flexural and impact properties, with 1.0 wt% giving the highest gains. Functional groups such as COOH and Si–OH improved bonding with the epoxy matrix and restricted CNT mobility. Silane treatment also improved CNT dispersion, resulting in stronger fibre–matrix adhesion. But physically treated CNT composites performed poorly due to agglomeration. The study concluded that while CNTs are effective nano-fillers, proper surface modification is essential to achieve uniform dispersion and significant mechanical improvement.

Anand et al. (2018) discussed the growing problem of non-recyclable polymer waste and suggested natural fibre–reinforced polymers as a sustainable solution. They developed epoxy composites reinforced with chemically treated kenaf and jute fibres. These treated composites showed much better mechanical and physical properties than untreated ones. Chemical treatment removed impurities such as pectin and hemicellulose, increased cellulose content and created a rough fibre surface. This improved bonding with the epoxy matrix and allowed better stress transfer. Water absorption was also reduced. Among all samples, the KHHK hybrid composite showed the best mechanical performance. Morphological analysis confirmed strong fibre–matrix interaction. Overall, the study showed that properly treated natural fibres can effectively reinforce epoxy composites for automotive and construction applications.

Zakaria et al. (2019) studied the effect of thermal cycling on injection-moulded polymer composites. Repeated heating and cooling reduced tensile strength, Young's modulus and elongation at break in all samples. This indicated a loss of stiffness and ductility. Dimensional changes were also observed. PP and kenaf/PP composites showed a reduction in hole size, while kenaf/MWCNT/PP and kenaf/MWCNT/MAPP/PP composites showed an increase. Among all materials, the kenaf/MWCNT/MAPP/PP composite exhibited the highest hardness. Thermal cycling also changed the fracture surface, leading to more brittle failure. The results confirmed that thermal cycling can significantly degrade mechanical performance and alter composite microstructure.

Napisah et al. (2021) investigated the mechanical behaviour of kenaf fibre composites hybridised with glass fibres and nanoclay. The addition of glass fibres significantly improved the strength of kenaf fibre reinforced polymer composites. Adding nanoclay further enhanced the mechanical properties of KFRP, GKFRP and GFRP composites. However, excessive nanoclay content at 5 wt% reduced performance due to particle agglomeration. Agglomeration increased resin viscosity and created stress concentration sites, leading to early failure. Tensile strength was higher in the longitudinal (0°) direction than in the transverse (90°) direction because fibres carried the applied load more effectively. Overall, the study showed that hybridisation and controlled nanoclay addition can improve composite strength, but optimal filler content is essential for reliable performance.

Khan et al. (2019) studied the effect of stacking sequence on the mechanical behaviour of kenaf–jute hybrid composites. Various laminate arrangements were fabricated using the hand lay-up method. Tensile and flexural properties were evaluated for each configuration. The J/K/J and K/J/K stacking sequences showed better mechanical performance than the other laminates. Among them, the K/J/K laminate achieved the highest tensile strength, exceeding both neat epoxy and the J/K/J laminate. Flexural strength and stiffness were also higher in the K/J/K configuration. Microscopic analysis showed fewer defects, less fibre pull-out and stronger fibre–matrix bonding in the K/J/K laminates. SEM images revealed voids, resin cracks, fibre bending and interfacial debonding, which supported the mechanical test results. Overall, the study confirmed that an optimized stacking sequence significantly improves composite performance.

Moyo et al. (2020) developed a nanoclay-reinforced kenaf/PLA biocomposite for impact-resistant applications. The addition of nanoclay improved medium-velocity impact behaviour by increasing energy absorption and resistance to cracking. These improvements were due to higher crystallinity, better fibre–matrix bonding and improved stress transfer. Performance increased with nanoclay content up to 5 wt%. Beyond this level, particle agglomeration reduced the benefits. At the optimal nanoclay content, the composite showed better perforation resistance and damage tolerance. The results indicate that kenaf/PLA nanocomposites are promising biodegradable materials for impact cushioning applications.

Sanjay et al. (2018) investigated jute/kenaf/E-glass hybrid epoxy laminates fabricated using the vacuum bagging method. The process produced laminates with low void content and uniform quality. Tensile behaviour depended mainly on the strength and stiffness of the fibres. Kenaf–E-glass hybrids performed better than jute–E-glass hybrids, but both were weaker than pure E-glass composites. Mechanical properties such as tensile strength, flexural strength and hardness were strongly influenced by ply arrangement. Laminates with glass and kenaf layers on the outer surfaces and jute in the core showed the best performance. Overall, these hybrid laminates offer a low-cost and biodegradable alternative for construction and household applications such as panels, roofing sheets, furniture and similar products.

Sathish et al. (2018) studied the mechanical and thermal behavior of banana–kenaf hybrid composites. The results showed that these composites are biodegradable and cost-effective. Among all samples, the composite with 60% kenaf (Sample 3) performed the best. It showed the highest tensile modulus and superior flexural and double-shear strength. This sample could withstand higher loads and absorb more energy than the others. Because of its good performance, Sample 3 was also tested under compression. The results showed that increasing thickness improved its load-bearing capacity. Thermal studies confirmed stable thermal expansion and better flame resistance. SEM images revealed that fibre stacking played an important role, with fibre pull-out and breakage explaining crack growth. Overall, Sample 3 showed balanced mechanical and thermal properties and was the most promising hybrid composite.

Iqbal et al. (2018) investigated UHMWPE/HDPE polymer blends reinforced with kenaf, basalt and hybrid kenaf–basalt fibres. Fibre reinforcement significantly improved tensile strength compared to the unreinforced blend. However, kenaf-reinforced composites showed slightly lower flexural and impact strength. The study confirmed that the UHMWPE/HDPE matrix is suitable for natural fibre reinforcement in load-bearing applications. HDPE also acted as a processing aid and improved compatibility with UHMWPE. The addition of kenaf, basalt and hybrid fibres expanded the application potential of these composites. Their balanced strength and easy processing make them suitable for applications such as biomedical components where very high stiffness is not required. Overall, hybrid natural–mineral fibre reinforcement allows tailoring of mechanical properties for specific needs.

Siti Nor et al. (2018) optimized twill-woven kenaf fibre composites using the Taguchi method. This approach helped identify the best fibre orientation with fewer experiments. Tensile strength and Young's modulus were evaluated using Minitab software. The optimized fibre angles showed higher stiffness and strength than non-optimized layouts. Experimental verification confirmed the Taguchi predictions. The study showed that fibre orientation strongly controls composite performance. Interactions between fibre layers also affected tensile and modulus values. Overall, the Taguchi method proved effective in improving the mechanical performance of woven kenaf composites.

Girimurugan et al., (2023) confirmed that biodegradable composites can be effectively produced using kenaf fibre (15–30 wt%) and kapok seed powder (5–20 wt%) in an epoxy matrix. Proper blending of these materials enhanced tensile, flexural, impact and elongation properties of the composites. The best tensile performance was obtained with 25 wt% kenaf fibre combined with at least 10 wt% kapok seed powder. Tensile strengths ranged from 28.23 MPa at 30 wt% fibre loading to 18.47 MPa at 20 wt%. Flexural strength and modulus also improved notably with 10 wt% kapok powder incorporated into the kenaf–epoxy system. However, mechanical properties declined when fibre content exceeded 25 wt% or filler content surpassed 10 wt%, suggesting an optimal range for effective reinforcement. Compressive strength and swelling behaviour were found to reduce proportionally with increased filler loading, with the highest compressive strength recorded at 18.67 MPa for 30 wt% fibre and 5 wt% filler. The reduction in performance was due to denser filler interactions, which weakened load transfer. Weak filler–matrix bonding also led to higher water absorption at higher filler levels. Since epoxy is widely used in structural applications, this composite shows good potential, including for future 3D printing applications.

Babu et al. (2023) studied kenaf–epoxy composites reinforced with powdered chick eggshell (PCES) particles. Processing lowered calcium content and increased oxygen in the eggshells. Uncalcined eggshells had higher calcium carbonate, while calcined ones showed less. Both types improved mechanical properties, but in different ways. Composites with uncalcined PCES had higher tensile modulus, flexural strength and hardness. Calcined PCES improved impact strength and wear resistance. The best values recorded were 49.57 MPa (tensile strength), 3.64 GPa (tensile modulus), 34.41 MPa (flexural strength), 3.49 GPa (flexural modulus), 16.43 kJ/m^2 (impact strength) and 63.91 HS (hardness). Optimal wear performance was at 2 wt% calcined PCES, while uncalcined particles performed best at 6 wt% and 2 wt% loadings. Overall, PCES fillers enhanced the key mechanical properties of kenaf–epoxy composites.

Saba et al. (2016) showed that adding nano oil palm empty fruit bunch (OPEFB), montmorillonite (MMT) and organically modified montmorillonite (OMMT) nanoclays improved the dynamic behavior of kenaf–epoxy composites. Nano OPEFB increased storage and loss modulus due to higher stiffness and hardness. The hybrid OPEFB–kenaf–epoxy composites also had lower damping, indicating restricted molecular mobility and better filler–matrix bonding. Glass transition temperatures improved for all hybrid systems. The improvement with nano OPEFB was similar to MMT and slightly lower than OMMT. Cole–Cole plots showed imperfect semicircles, confirming good fiber matrix interaction and material heterogeneity. These eco-friendly hybrid composites showed strong dynamic modulus and damping. It make them suitable for lightweight structural applications, including aerospace, building panels, marine structures, reservoirs and pressure vessels.

Norkhairunnisa et al. (2021) found that clay addition and fiber treatment strongly affect the thermal behavior of glass/kenaf/epoxy hybrid composites. Thermogravimetric analysis (TGA) showed that combining glass and kenaf fibers improved thermal resistance. Glass fibers provide better heat tolerance, while fiber–matrix interaction is enhanced. Adding clay further strengthened thermal stability by forming protective carbonaceous char. Treating kenaf fibers increased stability by removing lignin and hemicellulose, which delayed degradation. Dynamic mechanical analysis (DMA) revealed clear trends in storage modulus, loss modulus and damping behavior. The TKG1% composite had the highest storage modulus, showing increased stiffness due to nanoclay restricting polymer chain movement. It also showed the highest loss modulus from the combined stiffening effect of glass–kenaf hybridization. However, treated kenaf composites had the highest tan delta peak, reflecting reduced structural rigidity from hemicellulose removal. Overall, clay dispersion, fiber treatment and hybrid reinforcement together improved the thermal and viscoelastic performance of the composites.

2.5 IDENTIFYING THE EXISTING RESEARCH GAPS FOR THIS STUDY

A comprehensive review of the above literatures highlighted the following gaps in the existing research,

1. From the extensive literature survey, it was found that no studies have been conducted on turtle shell powder as filler.
2. Very few papers explore the potential of natural waste materials as eco-friendly fillers in high-performance composites.
3. No authors assessed the effect of turtle shell powder on tensile, flexural, impact and hardness properties of basalt/epoxy and kenaf/epoxy FRP composites to optimize mechanical performance.

4. Investigating particle dispersion and interaction with the epoxy matrix to improve composite design for engineering applications not much concentrated.

5. It is necessary to fabricate cost-effective composite materials by utilizing bio-waste fillers without compromising structural integrity.

6. From the literature survey it is clear that an exclusive study is required to bridge the gap between natural fiber composites and bio-waste filler utilization.

7. Clear Understanding of morphological characteristics and filler-matrix interaction is required to optimize composite structure.

2.6 MAIN OBJECTIVES OF PRESENT RESEARCH WORK

Basalt FRP composites are widely used in engineering applications for their strength and durability. Also enhancing their mechanical properties using sustainable fillers can reduce cost and environmental impact. This research explores particle-reinforced FRP composites for advanced structural applications. Although kenaf fibers are eco-friendly reinforcements, their mechanical performance will further improve by incorporation of natural fillers. Incorporating turtle shell powder can improve mechanical properties and provide an environmentally sustainable solution.

The objectives of present research work are listed below,

1. To fabricate the basalt/epoxy and kenaf/epoxy fiber-reinforced polymer composites with Turtle Shell Powder (TSP) as a novel bio-filler material at different weight percentages.

2. To analyze the influence of TSP on the mechanical properties such as tensile, flexural, impact strength, hardness, and density of basalt-epoxy laminates prepared using hand lay-up followed by compression molding.

3. To investigate the effect of TSP addition on the mechanical properties of kenaf-epoxy composites and compare the performance across different filler loadings.

4. To investigate the morphological characteristics and elemental composition of the TSP-filled fabrication of composites using SEM and EDAX to understand filler–matrix interaction, microstructural changes, filler dispersion, interfacial bonding, and fracture behavior.

5. To determine the optimal TSP content that enhance interfacial bonding while identifying conditions that lead to particle aggregation.

6. To evaluate the effectiveness of TSP filler, that how it impacts on overall composite performance.

2.7 DESCRIPTION OF THESIS WORK

Figure 2.1 shows the overall scheme of the present research. The study began with selecting the materials: epoxy as the matrix, basalt and kenaf fibers as reinforcements and turtle shell powder as a bio-filler for composite fabrication. The fabricated composites were subsequently characterized in terms of density, void content, tensile strength, flexural strength, impact resistance, hardness and microstructural features using SEM and EDAX analyses.

2.8 SUMMARY OF LITERATURE SURVEY

This chapter presents a comprehensive review of previous investigations on fillers used in kenaf FRP composites, basalt FRP composites and polymer matrix composites. It outlines the key research gaps and defines the objectives of the present study. Additionally, the chapter provides a schematic overview of the study's structure.

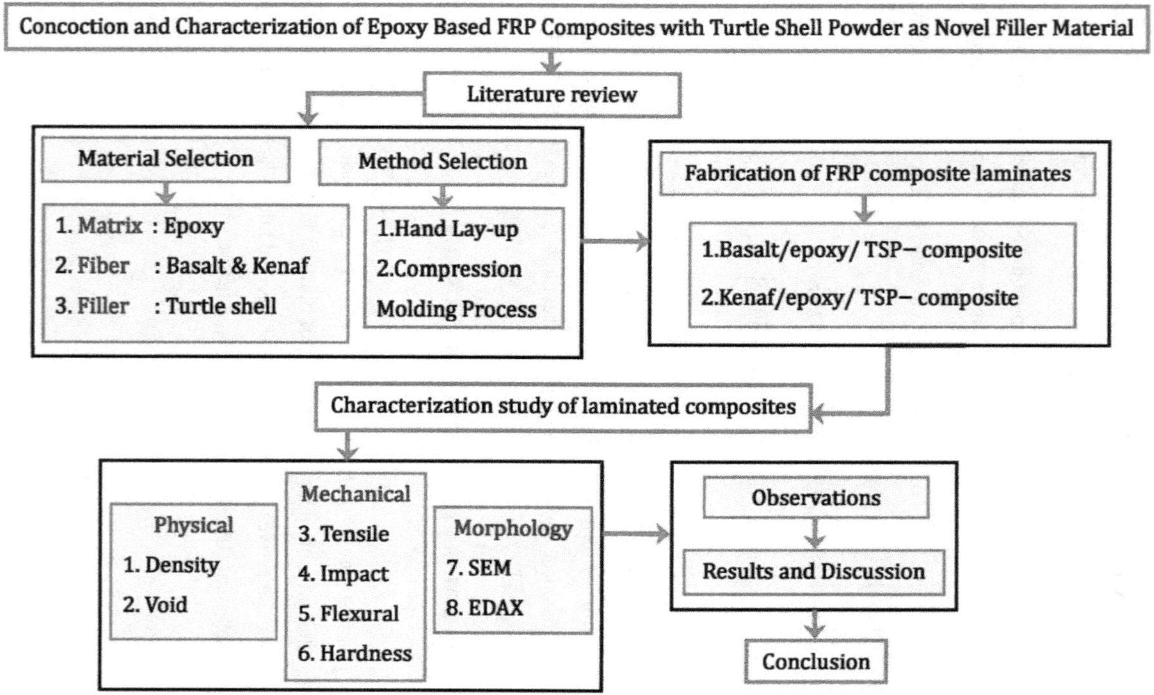

Figure 2.1 Flowchart of research work

CHAPTER 3

METHODOLOGY

3.1 Introduction

This chapter presents the selection of materials and fabrication procedures adopted for developing the basalt-epoxy composite. Comprehensive characterisation was carried out through density, tensile, flexural, impact and hardness measurements, along with SEM analysis, to evaluate the structural and mechanical performance of the newly developed composite material.

3.2 Materials

3.2.1 Filler Material

Turtle Shell Powder (TSP) is used as a filler material. The powder form of turtle shell was purchased from the E-Market of Alibaba.com. Table 3.1 and Table 3.2 show the chemical elements and properties of TSP. Figure 3.1 depicts the Turtle shell powder. Figure 3.2 shows the initial characterization of the TSP raw material.

Figure 3.1 Turtle shell powder

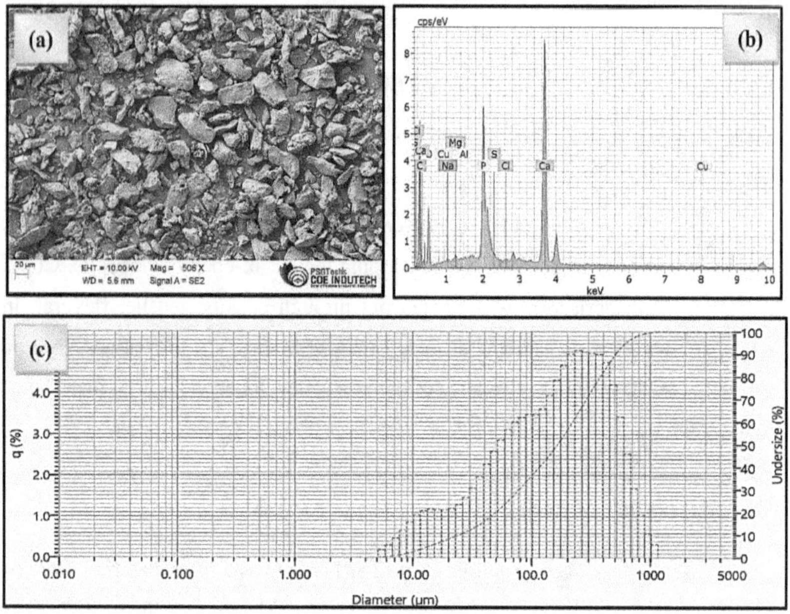

Figure 3.2 TSP (A) FESEM image, (B) EDAX results, (C) Particle size distribution.

Table 3.1 Chemical elements in TSP.

Element	Weight (%)
C	40.96
O	28.52
P	7.27
Ca	21.91
Cr	0.09
Na	0.24
S	0.16
Al	0.08
Mg	0.23
Cu	0.56

Table 3.2 Properties of Turtle Shell Powder [Zhang et al., (2012)].

Properties	Unit	Values
Density	g/cc	1.52
Tensile Strength	MPa	66.97
Elongation	%	2.06
Youngs Modulus	GPa	3.50

3.2.2 Fiber Material

Reinforcement-Basalt fiber bidirectional woven fabric weighing 210±10 GSM was supplied by Go-green Products, Chennai, India.

Table 3.3 Properties of Basalt fiber [Fiore et al., (2015)]

Properties	Unit	Basalt Fiber
Density	g/cc	2.8
Tensile Strength	MPa	2825
Elongation	%	3.15
Youngs Modulus	GPa	89

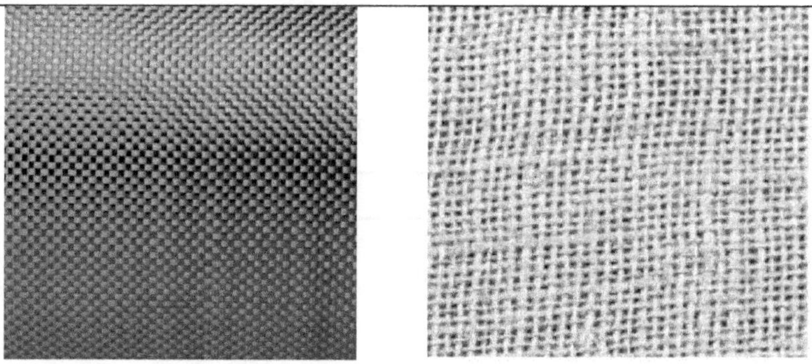

(a) (b)

Figure 3.3 Bidirectional woven mat - a. Basalt Fiber, b. Kenaf Fiber

The properties of reinforcement and matrices are shown in Table 3.3 and Table 3.4 shows the chemical elements present in the Basalt fiber. Figure 3.3 illustrates the Bidirectional woven mat of Fibers.

Table 3.4 Chemical elements in Basalt fiber [Fiore et al., (2015)]

Element	Weight (%)
SiO_2	42.43-55.69
Al_2O_3	14.21-17.97
CaO	7.43-8.88
MgO	4.06-9.45
Fe_2O_3	10.08-11.68
Na_2O	2.38-3.79
TiO_2	1.10-2.55
K_2O	1.06-2.33

Table 3.5 Properties of Kenaf Fiber [Ramesh et al., (2016)]

Material	Unit	Kenaf Fiber
Density	(g/cc)	1.4
Tensile Strength	(MPa)	284-800
Elongation	(%)	1.6
Youngs Modulus	(GPa)	21-60

Table 3.6 Chemical elements in Kenaf fiber [Ramesh et al., (2016)]

Element	Weight (%)
Cellulose	75.32
Hemi cellulose	21.5
Lignin	13.63
Pectin	3-5
Wax	0.28
Moisture	9.26
Ash	1.35

The kenaf fiber bidirectional woven mat was procured from Go-green products, Chennai, India. The properties of reinforcement are shown in Table 3.5. The chemical elements are shown in Table 3.6.

3.2.3 Matrix Material

Matrix –Araldite LY556 epoxy resin and Aradur HY951 hardener as an adhesive to prepare composite laminates, the corresponding resin and hardener were bought from Herenba Instruments & Engineers, Chennai, India. The properties of reinforcement and matrices are shown in Table 3.7. Figure 3.4 shows a snapshot of the matrix material.

Table 3.7 Properties of Matrix

Properties	Unit	Values
Density	g/cc	1.15
Tensile Strength	MPa	83
Elongation	%	4.2
Youngs Modulus	GPa	3.5

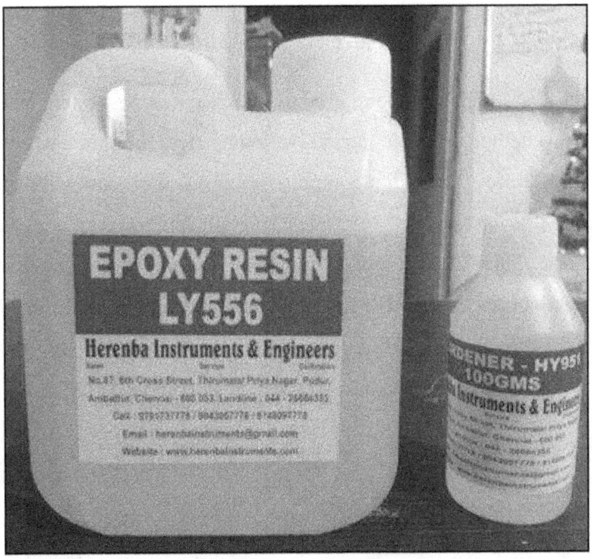

Figure 3.4 Matrix material

3.3 Fabrication Methods

3.3.1 Preparation of Matrix using TSP filler material

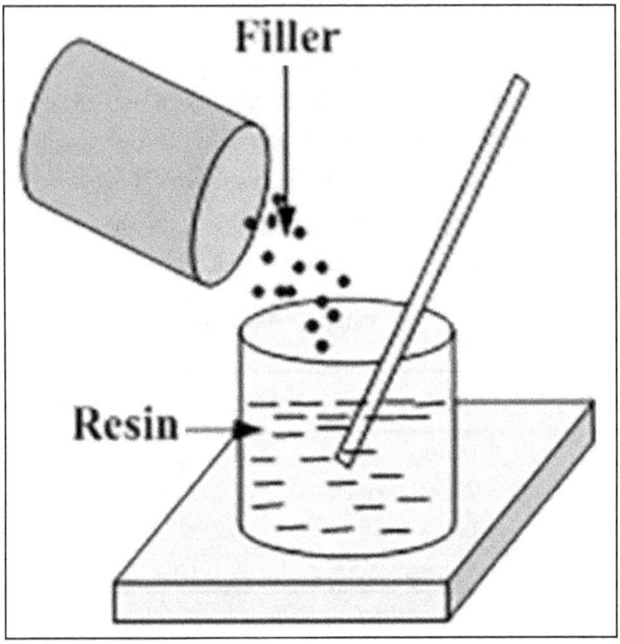

Figure 3.5 Mixing of Resin and Filler

The technique used to fabricate composite laminates was the hand-lay-up method. For the curing of composites, the laminates were placed in a hot mold press through a compression molding process with an applied pressure of 5MPa for 6 hrs at 80°C. The overall dimension of the final composite plate, 300×300×2.5±0.5 mm is obtained. For preparing the laminate, the stacking sequences of fiber (Basalt or Kenaf) were laid in multiple layers at different ply orientations, including quasi-isotropic $[0/90, +45/-45, 0/90, +45/-45, 0/90]_s$. Figure 3.5 illustrates the Mixing of Resin and Filler.

3.3.2 Preparation of Basalt / Epoxy / TSP-based FRP Composites

For preparing the Basalt laminate, the stacking sequences of basalt fiber laid 10 layers one by one at different ply orientations of quasi-isotropic

[0/90, +45/-45, 0/90, +45/-45, 0/90]$_s$ are as follows, while making the composite. The Turtle shell powder was added to the epoxy resin at different weight percentages of 1%, 3%, and 5%respectively, to the total weight of the matrix shown in Table 3.8. Then a stirrer was used to disperse the TSP evenly. The adhesive mixture of the epoxy resin and hardener was maintained in the weight ratio of 100:10 by parts. Furthermore, a good chain reaction and cross-linking of the matrix obtained were thoroughly stirred to get a homogeneous mixture. Figure 3.6 shows the fabricated composite using Basalt fiber.

Table 1.8 TSP proportion and code assigned for the laminated composites

Filler weight % of Basalt-epoxy composite laminates	**Code Assigned**	**Stacking Sequence [0/90°, +45/-45°, 0/90°, +45/-45°, 0/90°]$_s$**
Without TSP filled	B0	10 layers 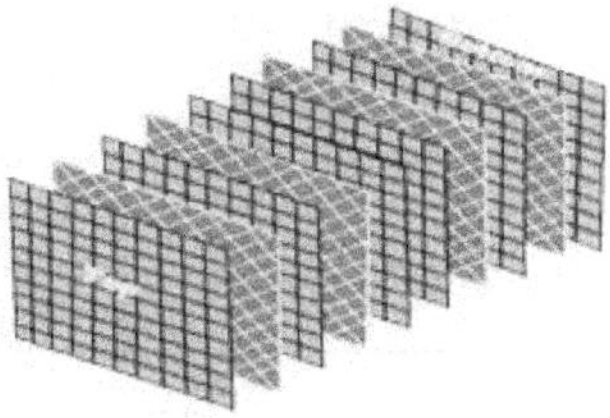
1 wt% TSP filled	B1	
3 wt% TSP filled	B3	
5wt% TSP filled	B5	

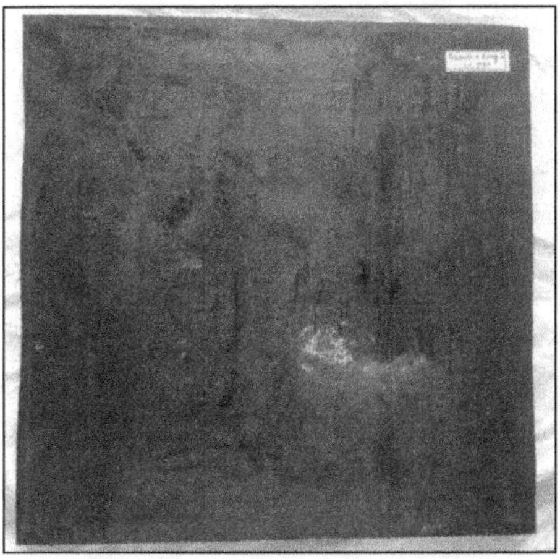

Figure 3.6 Fabricated composite using Basalt fiber

3.3.3 Preparation of Kenaf / Epoxy / TSP-based FRP Composites

For preparing the Kenaf laminate, the composite laminate consists of five layers with different ply orientations of quasi-isotropic [0/90, +45/-45, 0/90, +45/-45, 0/90] and is laid on a mold, as shown in Table 3.9.

Table 2.9 TSP filler percentage and code assigned for the laminates

Filler weight % of kenaf-epoxy composite laminates	Code Allocated	Stacking Sequence [0/90°, +45/-45°, 0/90°, +45/-45°, 0/90°]
Without TSP filled	K0	5 layers 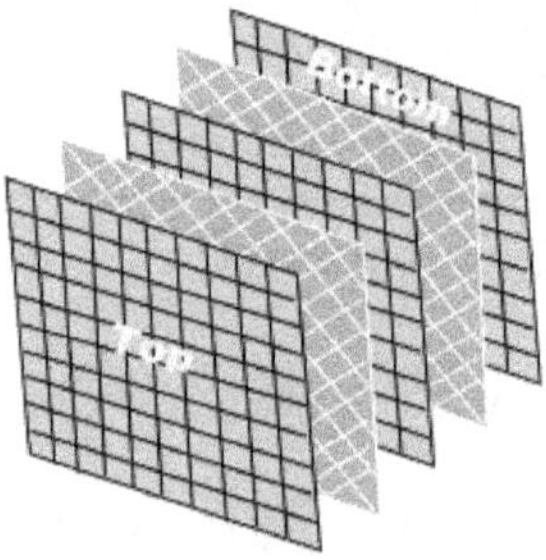
1 wt% TSP filled	K1	
3 wt% TSP filled	K3	
5wt% TSP filled	K5	

The weight ratio is 100:10 in parts that maintain the epoxy resin and hardener mixture. TSP was added to the epoxy resin at various weight percentages of 1%, 3%, and 5% of the total weight of the matrix, respectively. Since solid particles tend to clump and tangle when mixed, care should be taken to obtain a homogeneous laminated composite. The TSP was then uniformly distributed by using a stirrer. The weight proportion of the composites is shown in Table 4. Subsequently, the matrix mixture with TSP filler is poured and is squeezed on the layer-by-layer arrangement of fibers in the stainless-steel mold to develop a composite laminate. Kenaf/epoxy composite without filler was fabricated as a control sample. The final product of the composite laminate has a dimension of 300×300×3.5±0.5mm. Figures 3.7 and 3.8 show the fabricated composite using Kenaf fiber and the fabrication setup, respectively.

Figure 3.7 Fabricated composite using Kenaf fiber

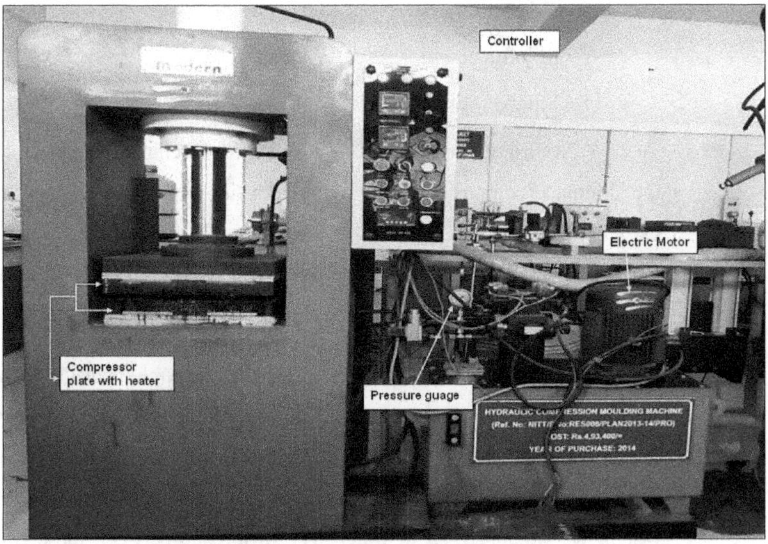

Figure 3.8 Hydraulic compression moulding machine for fabrication of composite

3.4 Experimental Techniques

3.4.1 Evaluation of Density

To determine the void content in the laminated composites, the theoretical density (ρ_{th}) and experimental density (ρ_{ex}) were calculated based on ASTM standard D2734. The void content % is determined by the following equation,

$$\text{Void content \%} = \frac{\rho_{th} - \rho_{ex}}{\rho_{th}} \times 100 \tag{3.1}$$

The theoretical composite density was calculated by considering each material's weight fraction and density using the following equation,

$$\text{Theoretical density } \rho_{th} = \frac{1}{\frac{W_m}{\rho_m} + \frac{W_r}{\rho_r} + \frac{W_f}{\rho_f}} \tag{3.2}$$

where,

W – Weight fraction

ρ – Density

m – Matrix

r – Reinforcement

f – Filler

According to ASTM D792 standard, Archimedes' Principle assessed the laminated composite's experimental density. Figure 3.9 shows the apparatus used in Archimedes' principle method, which is also known as the water immersion method. It is widely used due to its simplicity, accuracy and suitability for irregularly shaped specimens.

The water immersion method is based on Archimedes' principle, which states that a body submerged in a fluid experiences an upward buoyant force equal to the weight of the displaced fluid. In this technique, the density of a solid specimen is determined by measuring its weight in air and its apparent weight when submerged in water. The difference between these two readings corresponds to the buoyant force, enabling the calculation of volume and subsequently, the density.

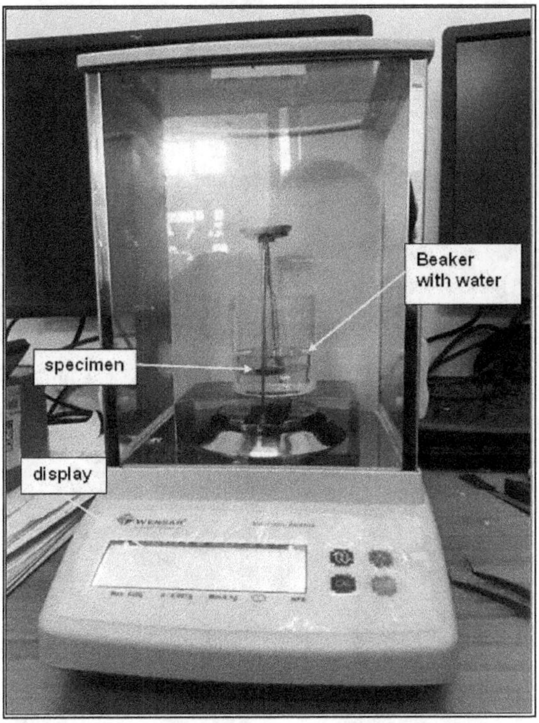

Figure 3.9 Density test - water immersion method test setup

3.4.2 Evaluation of Tensile Test

The Tinius Olsen HKL50 universal testing machine with a capacity of 50 kN was utilized to perform the standard ASTM D3039 for the measurement of tensile stress behaviour. The Tensile test setup is shown in Figure 3.10. The standard parameters were the strain rate of 0.01 min^{-1} and the constant displacement head speed rate of 2 mm/min. The specimen size required for testing is 250×25×3±0.5 mm, with a gauge length of 150 mm, which was cropped from fabricated composite laminates. The measurements of tensile stress, strain and modulus of elasticity data collections were evaluated and reported. Figure 3.11 shows the sample failure after the Tensile test.

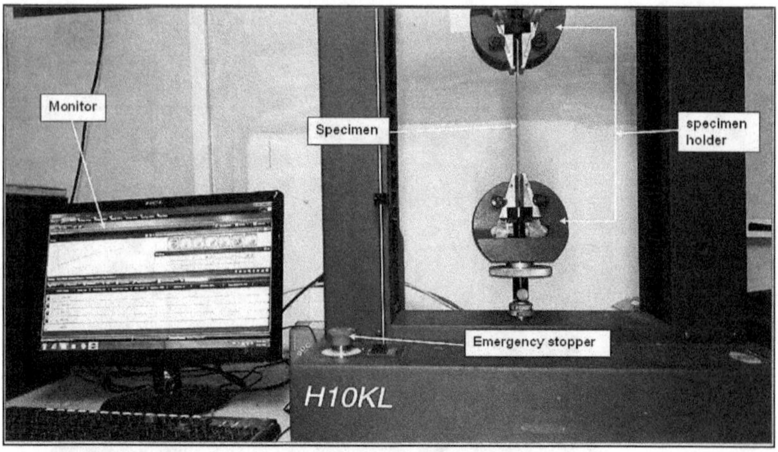

Figure 3.10 Tensile test setup

Figure 3.11 Sample failure after the Tensile test

3.4.3 Evaluation of Flexural Test

Steadiness and fortitude under different loading conditions are important to prompt the composites in various applications. Hence, the laminates have been experimented with the three-point bending technique was used in the flexural test performed in Tinius Olsen HKL50 universal testing machine as per ASTM D7264 standard. Figure 3.12 and 3.13 show the Flexural test and, test setup, respectively.

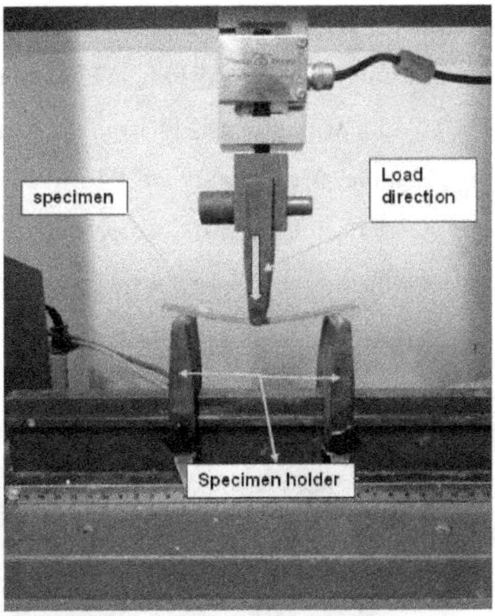

Figure 3.12 Flexural test

Figure 3.13 Flexural test setup

The required size of the specimen of span to thickness ratio is 32:1, the width is 13 mm and the specimen length is about 20 % longer than the support span was retrenched from the laminates. The crosshead speed movement of 1 mm/min was set. The response measured by this test is the modulus of elasticity, flexural stress, deflection strain and flexural stress-strain.

3.4.4 Evaluation of Impact test

The quantity of absorbed energy that measures the failure of laminated composites was investigated by using the Tinius Olsen (IT 504) Izod impact test machine. According to the ASTM D256 standard, the specimen size of $63.5 \times 13 \times 3 \pm 0.5$ mm was downsized from fabricated composites. A pendulum impacted the sample at a velocity of 3.46 m/s from a falling height of 609.6 mm. As per standard, the notch is made, which is used to initiate the crack to fracture the specimen. Figure 3.14 shows a view of the Impact test setup.

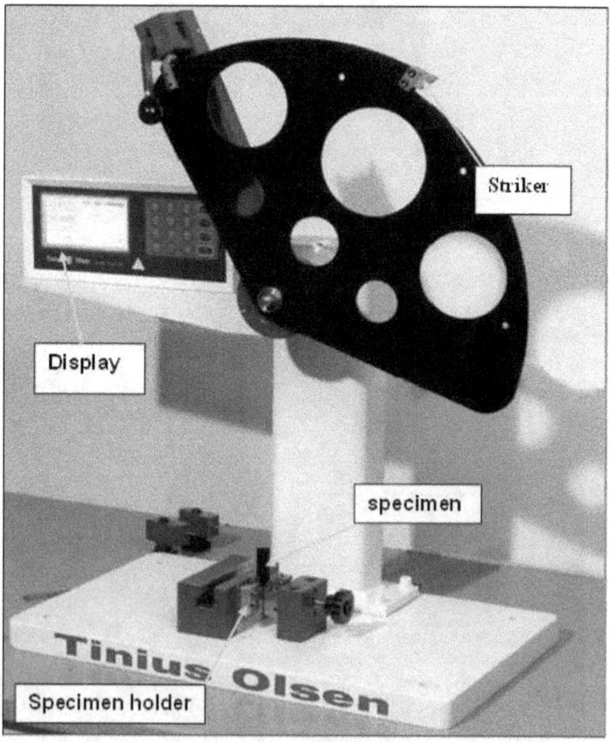

Figure 3.14 Impact test setup

3.4.5 Evaluation of Hardness test

The measurement of hardness is obtained by BarcolImpressor (Model No. 934-1), which is determined in accordance with ASTM D2583 standard test, where the surface resistance to indentation of a material is determined by plastic deformation. After each indentation, a minimum 3 mm distance is maintained from the indented point to the next indentation. A minimum of five readings were taken in each specimen. Figure 3.15 shows the instrument of the Hardness test setup.

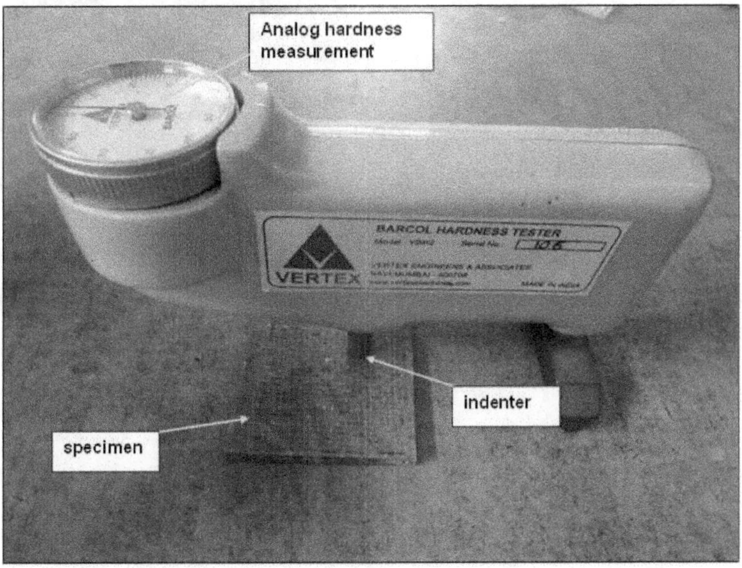

Figure 3.15 Hardness setup

3.4.6 Fracture analysis of fabricated composite by SEM

The composite laminates failed in the test experiments, then the fissured surface of the specimen was inspected and characterized to determine the behaviour of the material using a scanning electron microscope (SEM; Hitachi TM3000). The wrecked portion of the specimen depends on various factors like interfacial bonding, fiber orientation, delamination, etc., which were reported. The correlated failure mechanisms of TSP dispersion in composites were also examined. Before SEM analysis, the fractured surface specimen was coated with gold which sputtering covers by thin layer deposition on the specimen to prevent charging of specimen with electron beam. Figure 3.16 depicts the SEM setup.

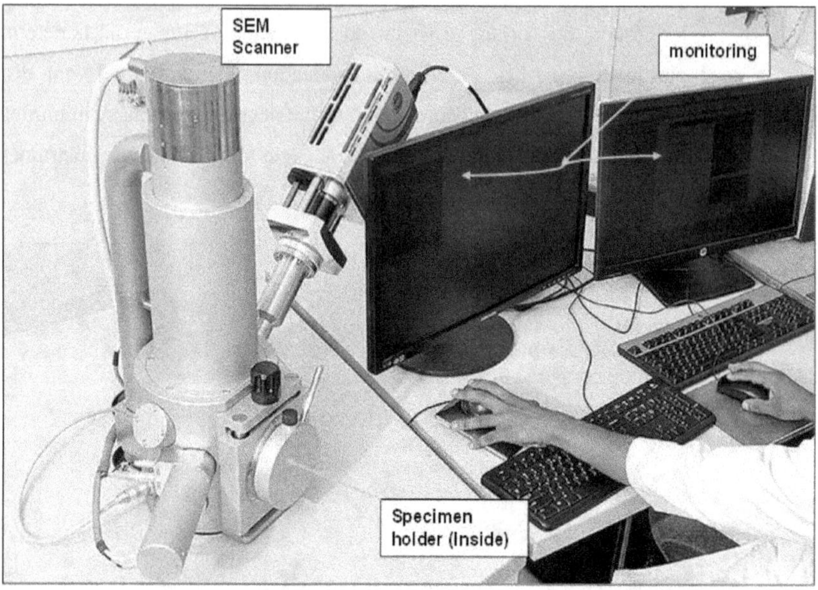

Figure 3.16 SEM setup

CHAPTER 4

CHARACTERISATION OF BASALT / EPOXY / TSP BASED FRP COMPOSITES

4.1 Introduction

The findings of researchers, helping to stimulate industrial production and foster innovative technologies. As environmental protection becomes an important research focus, many researchers are concentrating on the development of green composites due to their biodegrability, recyclability, renewability and overall sustainability. However, some constraints in industrial applications for the enhancement of material quality especially in defence applications we had been still used artificial materials. To save the environment, the niche material of natural fibers must be used to replace the synthetic fiber in the development of composites [Khan et al., (2018)]. Since last two decades, natural fibers have been surging in the replacement of artificial fibers due to lightweight, low cost, biodegradability and easily available material as an advantage of manufacturing polymer composites reinforcement. Nowadays, in aerospace, automotive and defence applications researchers focused on the viable usage of fibers from natural resources. Kenaf, hemp, jute, flax, bamboo and sisal, etc., are natural fibers which are used generally [Sanjay et al., (2018)].

Apart from these fibers, basalt fiber is one of the natural resources developed from basalt rock (i.e. volcanic rock). It is a prominent material that leads to an upsurge in the production of technological composite products in domestic and engineering applications, especially in defence area for the substitution of synthetic fibers in recent years [Fiore et al., (2015)]. Moreover, basalt fiber reserved the enticing excellent chemical, physical, mechanical, thermal, non-toxic and non-combustible properties. Based on these properties, basalt fibers have piqued the scientific community's interest has been used in FRPs composites [Regar et al., (2016)]. BF has good agreement with thermoset and thermoplastic materials compared to GF.

In thermoset polymers, epoxy resins are used broadly as matrix material in FRP composites thanks to having good corrosive resistance, excellent adhesion, curing at low shrinkage, compatibility, durability and availability [Vinay et al., (2021)]. [Vinay et al., (2021)] investigated the addition of Al_2O_3 nanofillers in basalt-epoxy composites, found that increase in flexural and ILSS upto 0.4wt%, impact and hardness increases upto 1.2 and 2.4wt% respectively and decreases in wear loss and tensile upto 2.4 and 0.6wt% respectively, further increase in filler wt% leads to the formation of aggregation of particles. [Arshad et al., (2021)] established the outcomes of TiC and coir fiber nanoparticles incorporated in basalt fiber reinforced bio/synthetic epoxy hybrid composites significantly increases in mechanical traits and higher thermal stability which enhances the highest load transfer capacity. [Pithalis et al., (2021)] aimed to fabricate, SiC as a micro filler in basalt-epoxy composite with different weight percentages disclosed the results of increasing the tensile strength by increasing the filler material, but the contribution of delamination failure decreases the flexural strength. The best result was found in the 7.5wt% composite. [Toorchi et al., (2020)] worked on the synergistic effect of graphene oxide and nanozirconia loaded in basalt-epoxy composites, the maximum bending strength and wear resistance gained in 0.1wt% GO+1wt% nano-Z as the results were exhibited. [Bulut et al., (2019)] experimentally investigated the characteristics of BFRP composites with nanoclay (NC) particles found that increases in tensile strength of 2wt%, a flexural strength of 1.5wt% and impact resistance of 0.5wt% about 7.61%, 29% and 16.8% higher compared to NC-0wt%, respectively.

[Matykiewicz et al., (2020)] studied the silanized basalt powder incorporated in basalt/flax hybrid FRP composite demonstrated that it significantly improved the mechanical and thermomechanical properties. [Jamali et al., (2019)] was carried out the study on FRP composite of basalt fiber and epoxy resin filled with graphene oxide (GO) and silanized graphene oxide (SGO) investigated on mechanical properties, the results revealed that the maximum tensile, flexural and compressive of strength and modulus obtained at 0.4wt% and 0.5wt% of SGO, respectively. Compared to GO, the SGO filler material is more efficient. [Li et al., (2019)] examined the epoxy/basalt fiber laminates with the incorporation of 2D montmorillonite (MMT), 0D nano SiO_2 and combined 2D MMT/0D nanoSiO_2 attained efficiently increased in ILSS and notch impact strength compared with plain epoxy/basalt fiber composites, the highest improvement in the incorporated composite of 2D MMT/0D nanoSiO_2 proved experimentally and had unique microstructure phase by SEM results. The mechanical behaviour of aluminium and basalt fiber epoxy FMLs composite in addition with modified nanoclay of 1, 3, & 5wt% experimental results showed that the Charpy impact absorbed the highest energy in 3wt% nanoclay content and decreased in ILSS due to the cluster of nanoclay particles studied by [Sambran et al., (2019)]. [Sampath et al., (2019)] evaluated the effect of moringa and bagasse ash filler particles added in epoxy/basalt composites exhibits better mechanical properties, corrosion resistance and water absorption resulting in moringa/basalt-epoxy composites. [Kim et al., (2018)] assessed the mechanical and thermal properties of natural graphite flakes filled basalt fiber reinforced epoxy composites, the results show that 20wt% of filler occurred an optimal performance of critical stress intensity factor, critical strain energy release rate, high thermal stability and conductivity enhanced significantly, also, the improvement in the properties due to crack pinning role mechanism performed in epoxy resin. [Mittal et al., (2018)] explored the reinforcement effect of CNTs were grafted on basalt fiber by the Chemical Vapour Deposition technique and for comparison study CNTs filled epoxy resin/basalt fiber composites were prepared. The output response indicated grafting of CNTs composites strongly improved the thermal and tribological properties. [Zhou et al., (2018)] followed the facile method by directly coating graphene on the surface of basalt fiber reinforcement polyamide 6 composites to investigate the mechanical and tribological properties which increased the tensile and flexural strength about 18.2% and 34% respectively and improved the storage modulus and glass transition temperature and also highly enhances the wear resistance.

[Bulut et al., (2017)] investigated the inclusion of graphene nano-pellets in basalt-epoxy laminated composite on mechanical properties such as tensile, flexural and impact tests with various loading content of 0, 0.1, 0.2 & 0.3 wt.%. The results indicated that the highest property obtained at 0.1wt% has resulted in high bonding strength of resin-fiber interactions. [Matykiewicz et al.,

(2017)] conducted the test of tensile, Charpy impact, DMTA and TGA on basalt fiber in addition to basalt powder reinforced epoxy composites at a different weight percentage of 0, 2.5, 5 & 10. The results concluded that the 2.5wt% composite combination expressed a better improvement in mechanical and thermomechanical properties, along with this, the degradation rate of an epoxy matrix is decreased due to the presence of basalt powder. [Manikandan et al., (2021)] experimentally worked on BFRP/epoxy composite with eggshell, water hyacinth and hybrid of both materials as bio-fillers at different proportions resulting that the hybrid bio-fillers are high-caliber than individual fillers improves the mechanical properties. [Jagadeesh et al., (2021)] incorporated the pongamia pinnata shell powder on sisal/kevlar hybrid composite by varying filler wt% and did the mechanical experimentation found that significantly enhances the composites.

[Ibrahem et al., (2021)] performed the wear test on chopped BFRP epoxy composites with the inclusion of sunflower seed husk, peanut shell husk and vegetable oils shows that efficiently improved the wear resistance. [Chowdari et al., (2021)] studied the mechanical characterization of areca fiber with coconut shell powder filled reinforced epoxy composites by varying the overall reinforcements wt% loading, they found that tensile, flexural and ILSS maximum at 30wt%, whereas impact and hardness obtained at 40wt%. [Gopal et al., (2019)] focused to make the composite added with seashell powder as filler on GFRP/epoxy at various compositions were investigated mechanically, which enhanced the tensile strength and hardness of the composite. The impact of crab shell particles filled sisal fiber composites was investigated by [Soundhar et al., (2019)] and they were discovered that adding the filler improves the mechanical performance upto 4wt%, further, adding a filler reduces the performance.

The aforementioned literary survey was conducted on the great achievements in the research of researchers, which shows that, to enhance the mechanical, thermal and tribological properties on FRP composites effectively. Hence, several and different types of organic and inorganic micro/nano filler materials are used with multiple proportions and hybrid combinations are accessed in composite production. In innovative thinking from the most researchers' knowledge, no one has tried using Turtle shell powder as a filler reinforcing material in BFRP composite fabrication. The current research work novelty is to examine the impact of Turtle shell powder as a filler-filled reinforcement in basalt-epoxy polymer composites fabrication and experimentally investigated the mechanical properties. The paramount material of Turtle shell powder used as filler material – Why? The uniqueness of the Turtle carapace formation presents macroscopic configuration of rigid boney ribs connected with sutures. The sutures are consisting of fibril reinforced keratin in the form of microscale layers consist of nano-fibrils and nano-platelets contain minerals are dispersed in keratin. Thus, a functional bio-shielding element of Turtle shell possess excellent mechanical properties acts as natural armour [Zhang et al., (2012), Ampaw et al., (2019), Shelef et al., (2017), Achrai et al., (2017)]. Hence, from the engineering science view point, the Turtle shell inspired to use as a filler material in the form of powder particles to improve the mechanical properties of the FRP composite.

Ballistic shields are layered composites that are commonly described as having a hard layer such as ceramics and a soft input. A projectile striking a hard plate can cause deformation and fragmentation of fragile armour, posing a risk to users [Nurazzi et al., (2021)]. Composites made of polymer and natural fibre have become widely used in bulletproof and ballistic applications. BF is a relatively new material and has lesser weight and high impact resistance. Due to this nature, the BFRP composites widely used for lightweight armors, missile packing box, and blast-resistant door fields, etc. According to several studies [Dehkordi et al., (2010), Nisini et al., (2017), Velumayil et al., (2022)], basalt fiber material used in the ballistic application for lightweight armors. The present study act as a frame work for the armor manufacturing and potential audience are researchers and manufactures who pertaining to armor application. Based on the ballistic application review, the fabricated composite may be suggested to utilize for an armor application.

4.2 Effect of TSP in Density

The reported composite density has theoretically and experimentally shown in Table 4.1. From the result, the experimental density is lower than theoretical density could be identified. This is quite overt as the former is computed mathematically assuming flawless material conditions. Here, the experimental density increases with TSP filler due to higher dense phase. The density of B1 laminate was lower compared to B3 and B5. The B5 composite density is higher due to the high dense formation of scutes, which was made by a fibrous protein called keratin. With reinforcing the void content was found to improve upto 3wt%, because better mixing and adherence may make the material more tightly bonded. After that, there may not be quite enough matrix material to adhere all of the reinforcing locations. The void content of B1 is higher due to less TSP content, also another factor that affects the density was minimal mold intricate geometry and awkward mixing of reinforcement and matrix [Sharma et al., (2020)]. Poor fiber-matrix interface bonding and partial wetting out of the fibers by the matrix may contribute to the rise in a void in B0 composites. The least void content percentage is preferred for product design and fabrication. The similar trend was identified with the researcher Kumar M et.al., (2021)

Table 1.1 Density of Composites

Sample Code	Theoretical density, ρ_{th} (g/cc)	Experimental density, ρ_{ex} (g/cc)	Void Content, (%)
B0	1.293	1.249	3.40
B1	1.499	1.454	3.01
B3	1.788	1.767	1.17
B5	1.872	1.839	1.76

4.3 Effect of TSP in Tensile Behavior

The composite tensile strength is depending upon the stretching on both ends at the maximum applied force without failure. The average tensile strength and modulus were depicted in Figure 4.1. The tensile strength obtained for the composite samples were 143.67, 192, 285, and 224 MPa as a result of B0, B1, B3 and B5 respectively. The resilient interface interaction which improves the tensile stress was developed by the hybridization phase [Dhiman et al., (2020)]. 285 MPa and 5.274 GPa was the maximum tensile strength and modulus obtained from the B3 composite sample which is 98.4% and 68.9% higher than the B0 composite (Table 4.2). A composite sample of B3 endows quite good bonding between fibers, matrix and TSP. The tensile strength was obtained optimally for the composites due to the load applied in the fiber orientation angle, in 0/90° the tensile force applied parallelly to the 90° orientation of fabric because the length of the fiber is larger [Kumar et al., (2020)]. The inclusion of Turtle Shell Powder (TSP) fortified the composite that has increased the tensile strength by the increase in weight percent, arising from the homogenous dispersion and interfacial interplay of fiber-epoxy and strong interactions between TSP–epoxy powder for the efficient stress-transfer which increases mechanical properties was attributed [Ary et al., (2020)]. In the B5 composite, the tensile strength found to decrease after the 3% weight of TSP which forms the agglomeration effect leads to failure and also shows that an increase in TSP in the epoxy, increases the energy required to rupture the composite samples [Pithalis et al., (2020), Bulut et al., (2017)] also it enhances the ductile properties. The inclusion of 5 wt% TSP filler resulted in the creation of tiny pores, which converged to form cracks, causing the specimen to collapse [Dhiman et al., (2020)]. Figure 4.2 shows the plots of stress-strain relations of tensile, most of the composite samples show the characteristics of brittle nature. The plots of the stress-strain curve behave linearly, where the stress increased due to strain increases until failure has reached.

The organic and inorganic nanocomposite platelets are building blocks in TSP [Chen et al., (2017)], it has been the enrichment of tensile strength for the laminated composite but shows the brittle nature. Bio-fiber reinforced thin platelets with a thickness of ~ 28microns are covered in TSP shell, hence the optimized design of the structure and material acts as optimal load-bearing capacity [Chen et al., (2017)], such a bio-fibers of TSP platelets may follow the stress direction on the laminates which helps to resist the evolution cracking.

Even though the tensile strain of B1 composite is higher compared to B3 composite because the stiffness of the basalt fiber and epoxy matrix is higher in B1.

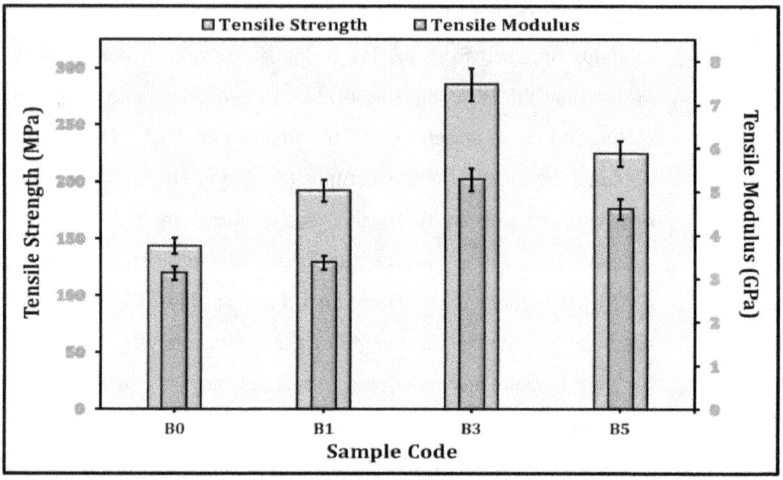

Figure 1.1 Average strength and modulus of TSP filler filled composites – Tensile

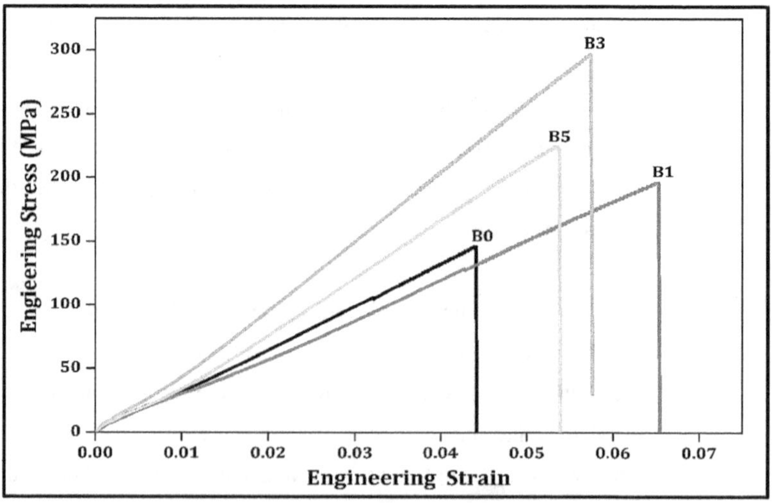

Figure 4.2 Emanated stress-strain performance– Tensile

Table 4.2 Mechanical Properties data of laminated composites – Experimented

Sample code	Properties					
	Tensile Strength (MPa)	Tensile Modulus (GPa)	Flexural Strength (MPa)	Flexural Modulus (GPa)	Impact Strength (J/m)	Hardness (Barcol)
B0	143	3.122	164	8.498	600	35.4
B1	192	3.369	224.67	11.123	722.22	40.0
B3	285	5.274	280	13.021	1061.11	46.2
B5	224	4.6	260	11.279	900	50.8

4.4 Effect of TSP in Flexural Behavior

The TSP behaviour effect of bending force on the laminated composite was conducted by flexural test.

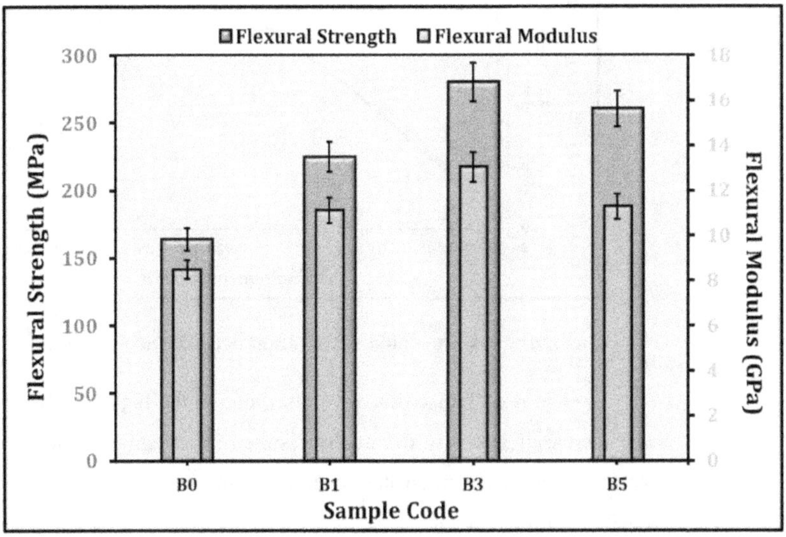

Figure 4.3 Average strength and modulus of TSP filler filled composites – Flexural

The average flexural stress and modulus were delineated in Figure 4.3 for the laminated composites of B0, B1, B3 & B5. The experimented results of flexural strength were obtained as 164, 224.67, 280 and 260.33 MPa for B0, B1, B3 & B5 respectively. The laminated composites of flexural strength and modulus are increased with TSP filler as far as 3%, then it is decreased in 5% filler content, otherwhere the similar results were reported with the inclusion of different filler [ASTM et al., (1987), ASTM et al., (2010), Sharma et al., (2020), Dhiman et al., (2020)]. The highest flexural strength and modulus were shown in B3 composite specimen which is 70.73% and 53.12% respectively higher than the B0 composite. The B5 laminated composite has 58.74% and 32.725% as strength and modulus respectively which is greater than B0, but smaller than the B3 composite (Table 4.2).

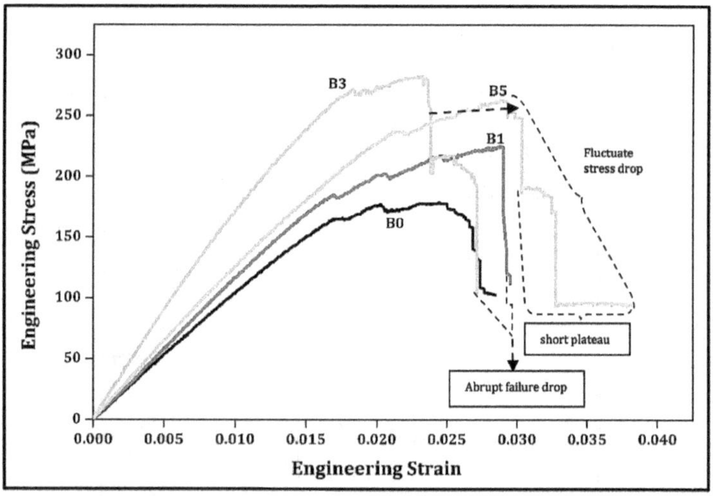

Figure 4.4 Emanated stress-strain performance– Flexural

The resin's viscosity was raised due to the high concentration of TSP. The increased viscosity eventually resulted in an agglomerated TSP structure which harmed the composites' overall performance [Sapiai et al., (2020)].The tensile, compressive and shear forces were evolved at the time of bending force applied in composite material [Pithalis et al., (2021)]. Moreover, the inclusion of TSP enhances the flexural properties which hold responsible for improving the bonding between fiber and matrix phase of the composite. The graphical flexural stress vs strain is illustrated in Figure 4.4. In this scenario, the flexural stress-strain of composites gradually rises for all investigated samples owing to the strong elongation characteristics of basalt fiber that shows good ductility property [Ary et al., (2014)]. The B0 & B1 samples showed typically elastic behaviour until reaches the damage or failure point. At that stage, the accumulation of cracks started to grow on the top of laminates, causing deviation of sudden failure drop behaves non-linearly which confirms the brittleness of composites.

For the samples, B3&B5 samples of the stress-strain curves partially behave as a linear elastic performance, upto the critical point reaches. At the failure point, the noted stress drop, which indicates fluctuate stress obtained due to delamination occurs on layer by layer will show the short plateau leads to an inexorable collapse in the laminate [Hashim et al., (2021)]. The load increased further or holds off as far as the second stage of failure point reached by reason of delamination in laminates. The crack takes it on oneself at the maximum load which affects the resin catastrophically in the composites. Thereupon, causing calamitous tensile failure at the bottom of the laminates. The results kickback until the abrupt failure of the entire laminated composite. The B3 sample has a steeper slope and reached high tensile strength over the other samples, owing to better adhesion in interfacial bonding between fiber and matrix. Moreover, the bio-fibres of TSP filler as a shielding material synthesize with good bonding between epoxy matrixes and enhance the strength of the composites. Finally, the B5 has shown intermediate responses, where the strain rate is maximum.

4.5 Effect of TSP in Impact Behavior

To determine the ability to break energy i.e. impact fracture resistance of the material, it was experimented with, by striking force using a weighed pendulum. Figure 4.5 exhibits the impact strength in the form of a bar chart. Based on the results, the noted impact strength of the laminates had the percentage increment is 20.37, 76.85, and 50 of B1, B3 and B5 samples respectively, compared to the B0 sample. But, the B5 sample decrements 15.18% when compared to B3 (See Table 4.2).

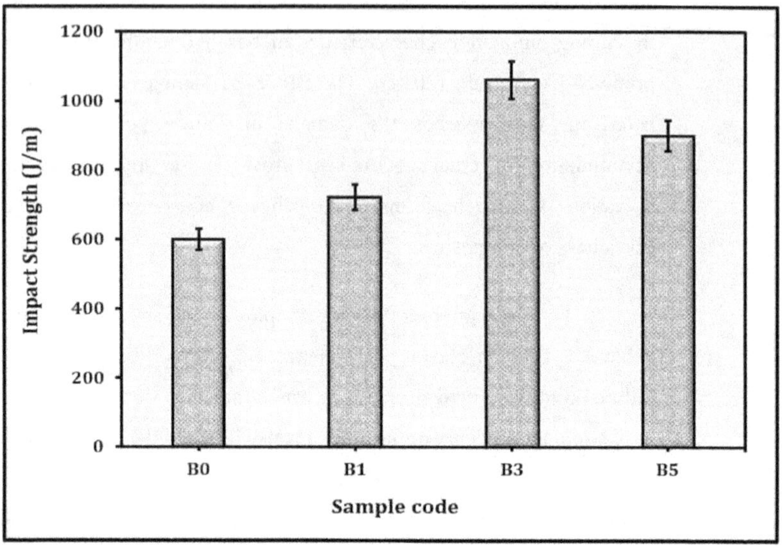

Figure 4.5 Emanated Impact performance – Izod

Because at the time of fabrication augmented viscosity face the moisture problem creates voids/pores within the composite forms weak bonding in B5 sample. A similar tendency was observed by [Darshan et al., (2020)], [Bulut et al., (2017)] and [Sharma et al., (2019)]. The breaking energy from the impact striker is maximum at B3 composite sample due to elastic deformation of fiber and intertwined/interlock nature of woven ply, matrix and TSP together which resists the impact load. Moreover, the orientation of the composite significantly improves the impact strength. The incorporation of TSP particles improves the impact strength on behalf of interface interlocking ability and excellent bonding of fiber and matrix, bestowing more energy absorption capacity, in conjunction with stronger and rigid properties compared with the B0 sample [Hashim et al., (2021)].

4.6 Effect of TSP in Hardness Behavior

The hardness value of the laminated composites by varying weight percentages of TSP particles is shown in Table 4.2. The results show that the hardness increases progressively with the inclusion of TSP filler increases as shown in Figure 4.6. The maximum hardness value obtained in the B5 sample of 50.8, which is 43.5% higher than the B0 sample.

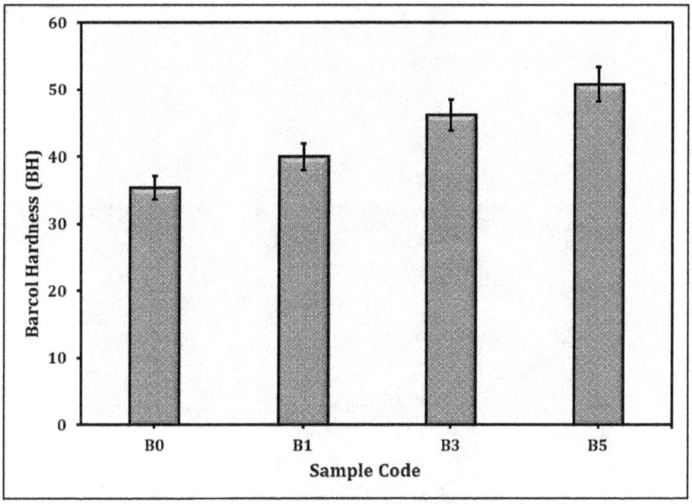

Figure 4.6 Emanated Hardness performance – Barcol

The hardness increases by reason of an increase bearing in load at higher particles which strongly resist the stress concentration on the laminate surface. Moreover, the elements present in the TSP particles provide more load resistance, and also good bonding with matrix material improves the integrity of ingredients leading to an increase in the hardness monotonically. The ductility of the fabricated composites is reduced as a consequence of the increased hardness, and the composites turn into brittle nature. Similar results have been discovered elsewhere with the admittance of various filler materials [Pithalis et al., (2021), Kumar et al., (2020), Sharma et al., (2019), Ary et al., (2014)].

4.7 Morphological Observations

The results of SEM were analysed for the fractured surface of tensile samples to examine the intrinsic failure of the materials. Figure 4.7 shows the SEM images of Tensile tested samples. The fracture analysis of the B0 sample was shown in Figure 4.7a and 4.7b, the fiber-matrix debonding leads to failure of delamination and fiber pull-out. The delamination failure occurs in 0/90° ply orientation because the load acts parallel to the fiber direction.

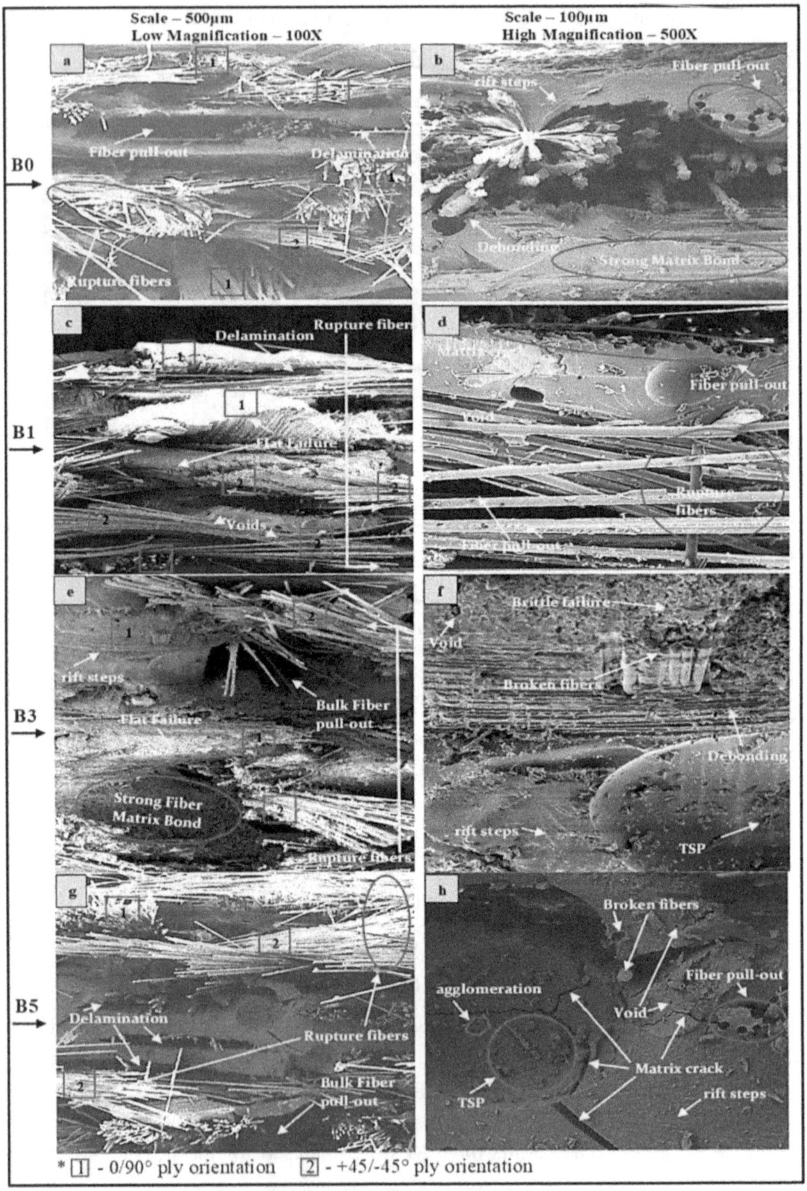

Figure 4.7 SEM analysis of fracture surface after tensile test of: (a, b) - B0, (c, d) - B1, (e, f) - B3,(g, h) - B5.

In Figure 4.7b, the cause of the strong interfacial fiber/epoxy bonding was shown to be the creation of rift steps formed in the epoxy matrix. Some failures that can be seen in the B1 sample are reinforcement/matrix delamination failure by lengthwise, debonding, fiber rupture and voids. B1 (Figure 4.7d) showed the expeditious crack propagation on the matrix owed by the formation of a void.

Most basalt fibre damage was found in the +45°/-45° ply orientation, and fiber/matrix delamination happened when the load was applied perpendicularly to the 0°/90° ply orientation. In the B3 sample, the coherence of filler and fiber-matrix materials were sturdy bond and load was observed. Due of the weak basalt-epoxy bonding, Figure 4e depicts the scanty of pull-out fibres. Also, we can view the broken fibers and interface reinforcement-matrix debond shown in Figure 4.7f, which determines the brittle failure of the sample in the presence of matrix crack and voids. Finally, the B5 samples of Figure 4.7g&h, exhibits the failures of delamination, debonding, bulk fiber pull-out, rift steps and fiber ruptures as discussed earlier.

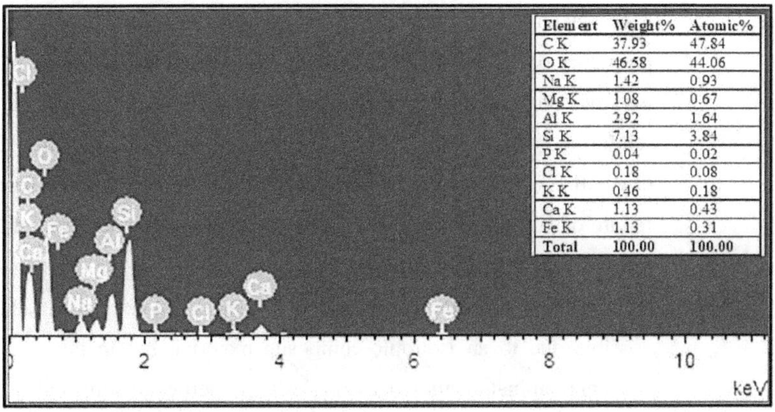

Figure 4.8 Elemental composition results ensure the presence of TSP after a tensile fracture of B5 specimen - EDAX.

Moreover, the appearance of matrix crack and fiber has broken, were raised by cause of agglomeration of TSP particle with the support of void was identified. In all the samples except B0, the incorporation TSP particle dispersed and good impregnation has been observed. Overall, the analysis of the fracture surface morphology points out the breakage of basalt fiber into different forms of short and long fiber pull-outs failure of 0/90° and +45/-45°respectively, which illustrates the brittle and ductile properties respectively. The presence of elemental components of TSP particles on the fractured surface of the tensile sample was confirmed with the EDAX result as shown in Figure 4.8.

4.8 Effect of TSP filler material and ply orientation

The TSP prevails the enrichment of composite to be strengthened, that explicit the strongest material which is attributed from the experimental results. The minerals present in the filler material have micro lamellas called keratin, employing combined with the mechanisms of flexibility and toughness had superior strength of the mineral [Chen et al., (2017)]. The filler enhances the mechanical properties, by the improvement of resistance of crack propagation on account of the distribution of biomineralized nanocomposite platelets. Alluring overall properties are conceded by the efficient micro and nanostructure of TSP filler. By adding this filler, the matrix builds up the nanostructure within the phase by altering interfacial interaction and bulk properties of the composites. The life utility of BFRP laminated composite has improved the stability and also cut down the failure of the matrix when it reinforces [Ary et al., (2014), Hashim et al., (2021)]. The effect of ply orientation improves the intrinsic properties which contribute to reducing the failure of laminated composites. The composite laminates are fabricated in the form of symmetrically arranged, that helps to boost up the bending stiffness and high impact strength. Furthermore, the active loads distributed evenly is possible due to an isotropic composite material, but in this case, the quasi-isotropic laminates mimic the composite properties to some extent [Agarwal et al., (2021)]. The fiber orientation angle enhances the strain of each layer when the load is applied in parallel and perpendicular direction of the laminates, that increases the tensile and bending strength respectively [Kumar et al., (2020)].

CHAPTER 5

CHARACTERISATION OF KENAF / EPOXY / TSP BASED FRP COMPOSITES

5.1 Introduction

The evolution of novel thinking mechanisms energizes scientists, leading to an active industrial manufacturing process and developing novel technologies. As there is a focus on environmental preservation, people often conduct studies on green composites. Its primarily motivated by their biodegradability, recyclability, renewability, and sustainability [Khan et al., (2018)]. The current scenario is a speedy growth in the usage of FRP composite materials. In recent years, significant research has focused on natural fiber-reinforced polymer composites due to their potential industrial applications and increasing consumer concerns regarding economic and environmental factors [Malik et al., (2021)]. Researchers emphasize the practical utilization of fibers derived from natural resources in aerospace, automotive, and defence applications. Natural fibers such as kenaf, hemp, jute, flax, bamboo, sisal and others find extensive application in various fields of engineering [Sanjay et al., (2018)].Materials researchers in green composites have shown significant interest in natural fibers derived from the kenaf plant (Hibiscus Cannabinus). It is primarily due to their exceptional properties, surpassing other natural fibers [Abdullah et al., (2011), Malik et al., (2023), Malik et al., (2024)].

As a natural fiber, kenaf is widely employed and has demonstrated substantial benefits across multiple industries, particularly in the automotive sector. It is readily accessible in large quantities and is a renewable resource. The fiber is biodegradable and sustainable, offering a low density. Consequently, it has been used in the development and manufacture of various thermoset and thermoplastic polymer composites [Saba et al., (2016); Raja et al., (2024)]. Kenaf fiber and its composites exhibit a specific stiffness comparable to that of glass fiber while priced at approximately two to three times lower than glass fibers [Malik et al., (2021), Anand et al., (2018), Taj et al., (2023)]. Epoxy is a thermosetting matrix or resin material containing one or more epoxide groups within its molecular structure. The majority of epoxy resins that are commercially accessible consist of oligomers derived from the di-glycidyl ether of bisphenol A (DGEBA).Like other thermoset resins, cured epoxy systems exhibit characteristics such as low fracture toughness, inherently limited impact resistance, retention of their shape after curing/polymerization, and decreased resistance to the initiation and propagation of cracks [Saba et al., (2016), Raj et al., (2024)]. [Saba et al., (2019)] investigated the addition of Magnesium Hydroxide (MH) reinforced as filler with various loading in kenaf/epoxy composite. They found that 20% MH shows better mechanical properties and thermal stability and stated that by increasing the filler percentage leads to decrease the properties. [Raj et al., (2022)] studied the effect of graphene filler in NaOH treated kenaf fiber-reinforced epoxy composites. They reported that increasing the graphene percentage increased the tensile, flexural, impact, and hardness. The 6% graphene-filled composite obtained the optimal results, whereas the 8% graphene filler showed good results in water absorption.

[Razak et al., (2019)] experimented on kenaf/PP with CNTs, MAPP filler as reinforcement before and after the thermal cycling process fabricated by injection molding. They found kenaf/MWCNT/MAPP/PP composite exhibited the highest hardness. Azizah et al. [Azizah et al., (2020)] investigated the mechanical properties and thermal behavior of silane treated on kenaf core reinforced polyurethane composites. The TMPS treated kenaf core fiber with 20% loading showed the most effective improvement. [Natrayan et al., (2022)] explored the effect of adding graphene fillers to kenaf fiber epoxy-based composites; the results show that 3% wt filler with salinized kenaf composite exhibited good mechanical performance as 16% higher than other composites. The mechanical, thermal and DMA analysis of kenaf/epoxy composites with a reinforced filler material of cellulose nano fibers and nano clays (MMT&OMMT) at 0.75wt% reveals the experimental result of enhanced all the properties for OMMT based composite was observed by [Khan et al., (2020)].

[Saba et al., (2016)] established the result of adding OMMT/kenaf/epoxy hybrid nanocomposites improved the density and thermomechanical properties. [Tay et al., (2012)] fabricated the kenaf/PP composites with filler material of montmorillonite (MMT) at various wt% and showed considerable improvement in flexural and impact properties; also, the addition of E-43P gives additional support to increase the performance rate of the composite properties. [Ismail et al., (2010)] studied the effect of filler loading and compatibilizer on kenaf core filled into HDPE/soya powder composites. They concluded that the tensile modulus, flexural strength & modulus and water absorption increase with increased filler material. However, tensile strength and elongation at break show a decreased trend; the MAPE also supports the tensile and flexural properties. [Din et al., (2010)] experimentally analyzed the composite of kenaf/ unsaturated polyester with UNT-MMT and CTAB-MMT filler and found that the tensile properties increased, the superior result performed by 3wt% of CTAB-MMT, which improves the effective distribution of filler and long polymer chain reaction within the matrix. [Taj et al., (2023)] investigated the effect of nano aluminium oxide and nanographene in kenaf/ epoxy composite. They demonstrated that adding filler material at 6% of nano aluminium oxide and 3% of nanographene increases the density and decreases the void content, lowering the water absorption.

[Lee et al., (2017)] incorporated the Magnesium hydroxide filler in kenaf/PP composites with varying fiber and filler weight percentage and experimented with the mechanical and thermal properties found that significantly improves the properties. [Kamarudin et al., (2018)] studied the mechanical and morphological properties of PLA based kenaf fiber composite with EJO vegetable oil and resulted that the 40% treated kenaf composite enhanced the mechanical properties. [Ismail et al., (2010)] Investigated the kenaf/halloysite nanotubes hybrid filler natural rubber compound, the results showed that the tensile strength, elongation at break and fatigue life was increased by increasing the HNT weight percentage, but tensile modulus decreases.

There are various organic and inorganic micro/nanofiller materials have been utilized in composite manufacturing, employing different proportions and hybrid combinations The turtle shell powder was chosen as the primary filler material because of the distinctive characteristics of Turtle carapace formation manifested in the visible arrangement of rigid boney ribs interconnected by sutures. The sutures comprise layers of microscale fibril-reinforced keratin, which contains dispersed minerals in the form of nano-fibrils and nanoplatelets. As a result, the turtle shell serves as a functional bio-shielding element with exceptional mechanical properties, functioning as a form of natural armor protection [Achrai et al., (2017), Shelef et al., (2017), Zhang et al., (2012), Ampaw et al., (2019)]. Therefore, the turtle shell filler material will use as powdered particles inspired to enhance the mechanical properties of the FRP composite.

5.2 Density and Void content

It's worth noting that density is an important property to consider when designing composites for specific applications. Higher-density composites can be stronger and stiffer but may also be more brittle and less impact-resistant. Lower-density composites may be more flexible and impact-resistant but less strong and stiff. Therefore, the choice of filler and its concentration should optimize to achieve the desired balance of properties for the specific application. In FRP composites, the natural fibers are used to increase the strength-to-weight ratio of the composites. Table 5.1 shows the theoretical density, experimental density and void content for the laminated K0, K1, K3 and K5 composites. The filler material of TSP increased the experimental density of the KFRP/epoxy composite. Also, it identified that the observed density is lower than the theoretical density. Overall, the K5 composite sample shows the highest density compared to the rest due to the formation of scutes, which consists of a fibrous protein called keratin. K1 (7.63%) exhibited a higher void content than K5 (1.81%), owing to the high porosity structure of the woven kenaf mat, which has fibers arranged more loosely. It leads to a significant amount of resin extruding from the mat during the hot press process.

Table 5.1 Density of Composites

Sample Code	Theoretical density ρ_{th} (g/cc)	Experimental density ρ_{ex} (g/cc)	Void content (%)
K0	1.273	1.169	8.17
K1	1.271	1.174	7.63
K3	1.269	1.189	6.30
K5	1.267	1.243	1.81

Moreover, the air entrapped in the interface of the filler matrix was encountered during the mixing process. Specifically, by the hand-layup method, in the composite laminate fabrication, still, the presence of void content is irresistible [Sand et al., (2020), Pawar et al., (2015)].

5.3 TSP behavior in Tensile analysis

The stretching on both ends when the maximum force is applied without producing any damage determines the tensile strength of the composite material. Figure 5.1 portrays the effect of TSP filler on the tensile strength and modulus of the composite. The composite samples achieved tensile strength values of 28.17, 32.60, 35.52 and 50.16 MPa, corresponding to K0, K1, K3 and K5, respectively. The K5 composite sample exhibited a maximum tensile strength of 50.16 MPa and a modulus of 2.905 GPa, representing a 78.06% and 97.89% increase, respectively, compared to the rest of the composite samples (Table 5.2).

Introducing the strong TSP filler into the epoxy matrix enhances the tensile modulus of composites. This enhancement is achieved by restricting the movement of the macromolecular polymeric chains [Saba et al., (2019)].

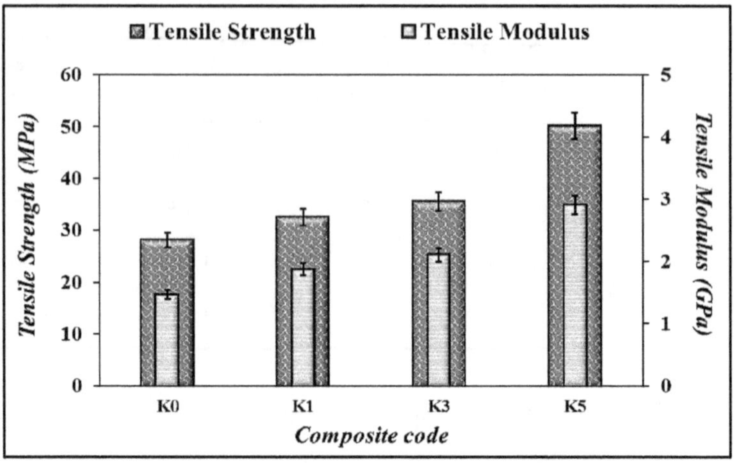

Figure 5.1 Strength and Modulus of TSP/kenaf/epoxy composite – Tensile

Table 5.2 Mechanical Properties of laminated composites – Experimented

Sample code	Properties					
	Tensile Strength (MPa)	**Tensile Modulus (GPa)**	**Flexural Strength (MPa)**	**Flexural Modulus (GPa)**	**Impact Strength (J/m)**	**Barcol Hardness (HB)**
K0	28.17	1.468	49.5	3.33	72.22	23
K1	32.60	1.876	53.2	3.53	77.77	26.4
K3	35.52	2.106	58.7	3.86	83.33	28.6
K5	50.16	2.905	71.1	4.61	100.2	34

The addition of TSP strengthened the composite, resulting in an increase in weight percentage that increased the tensile strength. Efficient stress transfer is achieved by ensuring uniform dispersion of fibers within the epoxy matrix and strong interactions between TSP-epoxy powder. These factors enhance the mechanical characteristics by promoting homogenous distribution and interfacial interactions between the fiber-epoxy components [Kasinathan et al., (2022)].

The composites achieved optimal tensile strength by applying the load at the 0° fiber orientation angle. In the 0/90° configuration, the tensile force is applied parallel to the 90° fabric orientation because the fiber's longer length contributes to increased strength [Salman et al., (2016)]. Figure 5.2 shows most of the composite samples exhibit a brittle nature trait, as indicated by the stress-strain plots of tensile testing. The stress-strain graphs behave linearly, increasing stress as strain increases until failure is attained. Incorporating organic and inorganic nanocomposite platelets in TSP contributes to the enhancement of tensile strength in laminated composites, yet it also results in a brittle nature displayed [Chen et al., (2017)]. TSP shell covers bio-fiber reinforced thin platelets with a thickness of 28microns; therefore, the structures and materials optimized design serve as the best possible load-bearing capacity [Zhang et al., (2012)].

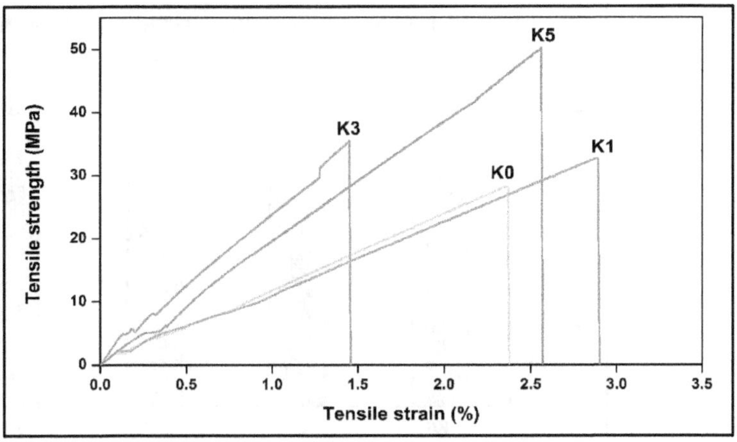

Figure 5.2 Emanated Tensile stress-strain performance

Due to the higher stiffness of the kenaf fiber and epoxy matrix in K1, the tensile strain of the K1 composite is greater than that of the K5 composite sample. The current findings are consistent with those of the other research article, which discovered that adding oil palm nano filler enhances both the tensile strength and the tensile modulus of kenaf/epoxy composites [Saba et al., (2016)].

5.4 TSP behavior in Flexural Behaviour

A composite material's bending strength is determined by flexural testing, which is done by a 3-point bending test. Figure 5.3 depicts the laminated composite's flexural strength and modulus for K0, K1, K3 and K5. The flexural strength values obtained for K0, K1, K3, and K5 were 49.5, 53.2, 58.7 and 71.1 MPa, respectively. The K5 composite specimen displayed the maximum flexural strength and modulus, which were 43.64% and 38.44% greater than the K0 composite, respectively (Table 5.2). The composite's higher stiffness by dint of the fiber and TSP additions prevented fracture formation early [Gurusamy et al., (2023)]. Hence, the composite's flexural strength is enriched. When TSP was added to the epoxy as filler, the composites transmitted tremendous stress along with the kenaf fibers.

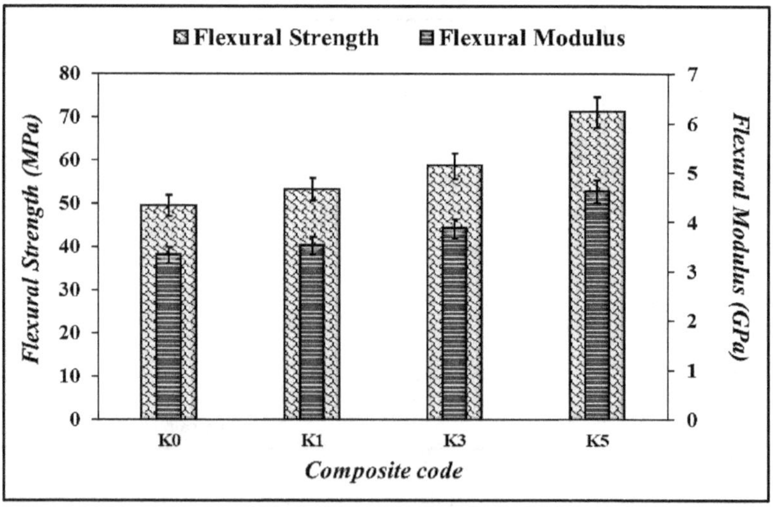

Figure 5.3 Strength and Modulus of TSP/kenaf/epoxy composite–Flexural

The high TSP content increased the viscosity of the resin. These enhancements are explained by the increased interface area, rigidity, and stiffness of the additional TSP filler, which allows the applied stress to communicate efficiently through improved interfacial adhesion [Saba et al., (2019)]. The composites' overall performance was negatively affected by forming an agglomerated TSP structure caused by the eventual increase in viscosity. Previously, comparable outcomes were reported using a different filler [Sapiai et al., (2020)]. When bending force was applied to a composite material, tensile, compressive and shear forces were generated [Pithalis et al., (2021)]. Figure 5.4 shows the graphical representation of flexural stress vs strain behavior. In this plot, due to the significant elongation features of kenaf fiber, which exhibits good ductility properties, the flexural stress-strain of composites steadily increases for all examined samples. The K5 laminated composite sample showed to withstand the highest loading capacity, which behaves partially as linear elastic performance until the yield point. The elastic behavior for all the samples was normal until they reached the damage or failure threshold. The flexural effect of the fiber yarns in the warp and weft directions can produce an interlocking structure in various orientations, which can restrict the fiber yarns' ability to extend along the directions and produce fibers with a higher bending load capacity [Salman et al., (2015)].

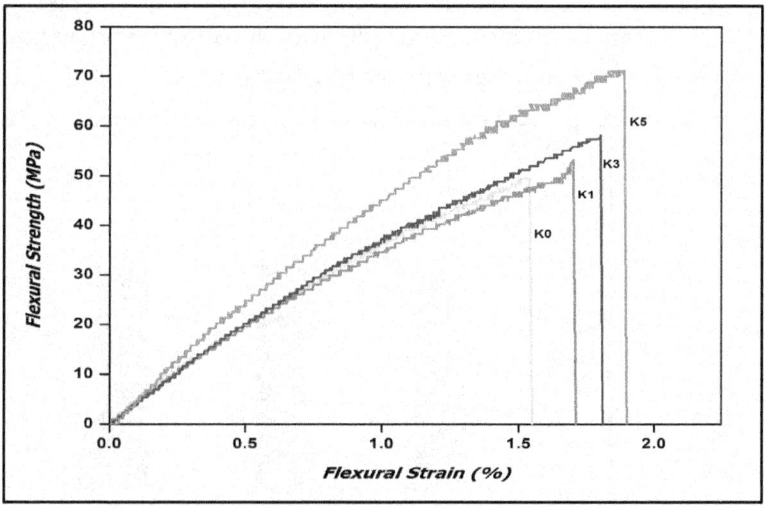

Figure 5.4 Emanated Flexural stress-strain performance

When the load is increased further, the failure occurs by delamination of layers when the resin affects catastrophically in the composites. With this, the identified additional reason that the TSP filler might be the development of weak spots leading to early rupture by the accumulation of the TSP particles, which caused the inciting interaction of particles that affects the flexural properties. According to research investigations, flexural qualities vary depending on the kind of polymer, and fiber orientation is crucial in determining mechanical capabilities [Salman et al., (2016), Sreekala et al., (2002)]. From the observed results, it can infer that the TSP filler in the composites based on kenaf/epoxy significantly enhances their mechanical properties.

5.5 TSP behavior in Impact analysis

The material was tested using a weighted impactor to apply force to establish its capacity to break energy or impact fracture resistance. Figure 5.5 presents the bar graph depicting the results of the impact strength experiment conducted on the TSP/kenaf/epoxy composite. According to the experimental results, the impact strength of the laminated composite for K0, K1, K3, and K5 is 72.22, 77.77, 83.33, and 100.2, respectively. From the results, the K5 composite sample has a maximum tensile strength of 97.89% higher than the K0 sample. After incorporating the turtle shell powder, a noticeable enhancement in the impact energy of the composite was observed.

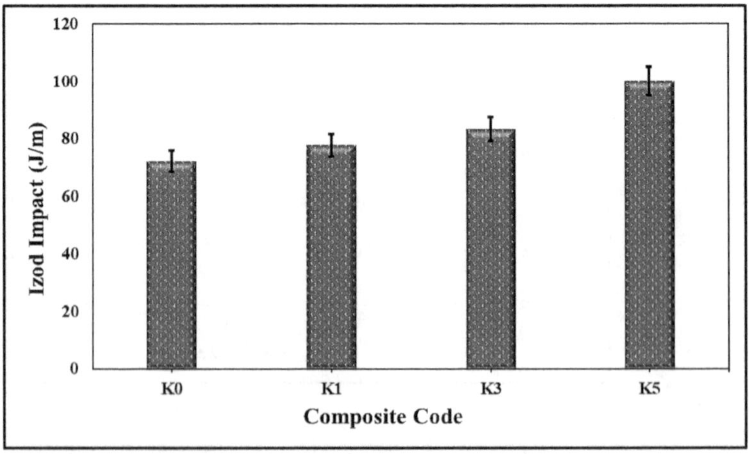

Figure 5.5 Emanated Izod Impact strength

Thus, TSP helped to absorb an amount of impact energy to some extent during the sudden impact load. As the percentage filler content increased, the propensity of composites' impact strength also raised. The K5 sample energy absorption is maximum due to inter bind and interlace of fiber layer, matrix and TSP filler in the composite to endure the impact load. Matrix/fiber fracture, matrix/fiber debonding, and pullout of fibers were the necessary factors considered for the impact test. When shear forces transfer the applied load to fibers, the load can surpass the strength of the bond between fiber and matrix, leading to debonding [Jawaid et al., (2018)]. The fracture spreads along the fibers' weft and warp directions, but nearby fibers may stop in both directions. The pendulum is parallel to the weft direction of 0/90° fiber layers; while it strikes, it is a chance for fibers to pullout in the composite. On the contrary, ±45° fiber layers decreased the crack propagation and fiber fissure, owing to the applied impact force transformed isotropically. Thus, the crack propagation was arrested by the support of the arrangements of fibers before the failure in K5 laminated composite [Manral et al., (2020)].

5.6 TSP behavior in Hardness analysis

The test will assess the material's ability to withstand penetration depth when subjected to a load. The hardness level will be higher in the case of brittle material. The experiment on Barcol Hardness for each sample is shown in Figure 5.6. The noted result shows that the 5wt% TSP filled in the K5 sample had the highest hardness value of 34 and the 0wt% TSP filled in the K0 samples showed least hardness value of 23. As the percentage of filler content increased, the hardness also increased gradually, which is 14.8%, 24.4% and 34% of K1, K3 and K5is higher than the K0 composite sample, respectively. The constituents within the TSP filler offer enhanced load resistance, bolstering the integrity of the components and resulting in an elevation in hardness and a good density of the interfacial bonding of matrix and filler. It suggests that the TSP particles applied a sufficient counterforce action against the impressor. Furthermore, the minuscule TS powder filled the gaps between the epoxy and fiber, reducing void volumes. Other studies using filler materials have had similar outcomes [Razak et al., (2019), Arun et al., (2019), Raj et al., (2022)].

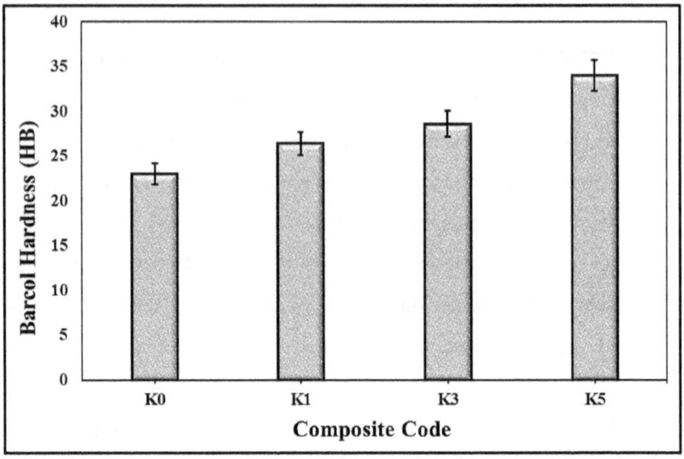

Figure 5.6. Emanated Barcol Hardness

5.7 Morphological analysis

The snapshot of Figure 5.7 shows the performed SEM analysis to examine and depict the morphological characteristics of the fractured surfaces of kenaf/epoxy composites without filler and with TSP composites. Figure 5.7 (A, B) identifies the presence of increased formation of voids, matrix cracks and debonding. The debonding of the fiber-matrix phase initiates the failure of delamination and fiber bundle pullout. Delamination failure is observed in the 0/90° ply due to the parallel application of the load along the direction of the fibers. The +45/-45° ply seems to fracture by combining matrix crack propagation, fiber pulling out, and fiber breaking. Furthermore, the K0 composite presents visual evidence of fiber breakup and the debonding between the reinforcement interface and matrix. These factors play a crucial role in causing the sample to fail in a brittle manner when matrix cracks and voids are present. The micrograph of the K5 composite in Figure 5.7 C reveals a complex route due to cracks as they propagate through the interface between the fiber and matrix. In addition, numerous empty spaces are visible following the fracture, suggesting that the bulk fiber pull-out phenomenon played a significant role. This analysis clearly illustrates that the effective dispersion and even distribution of TSP filler enhance the bonding between the polymer and kenaf fibers in the composites.

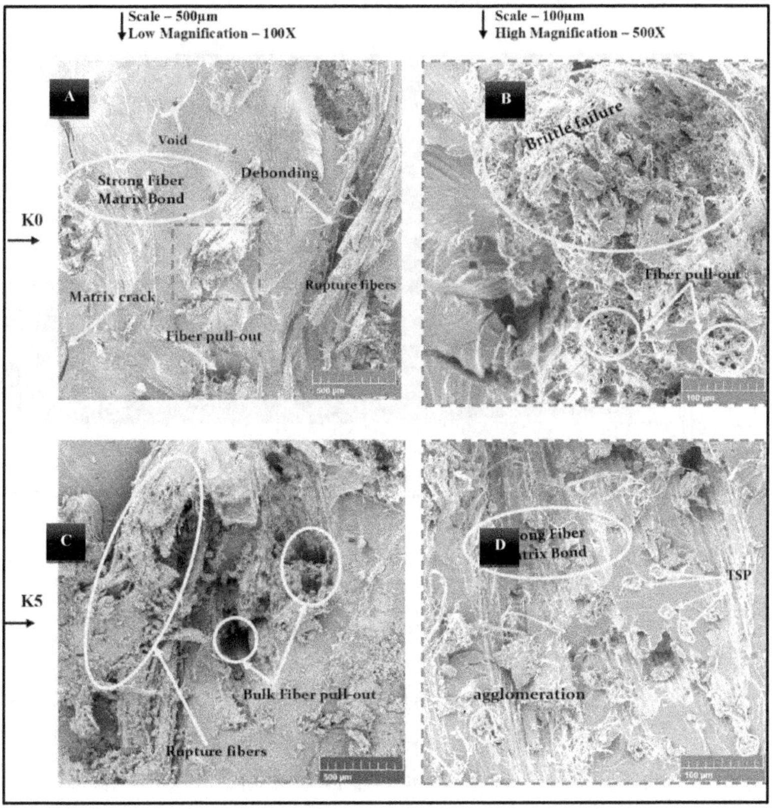

Figure 5.7 SEM analysis of tensile fracture failure of (A,B) – K0, (C,D) – K5

A strong bond between the TSP filler and fiber-matrix materials was evident, resulting in observed load consistency, as shown in Figure 5.7 D. Generally, the filler materials used in the FRP composites will form the agglomeration effect of nano or microparticles because of improper mixing of matrix and filler, size of the particles, and filler content etc. Likewise, the aggregation of TSP particles increased the visibility of matrix fractures and broken fibers, leading to the discovery of voids. As discussed previously, the failures like delamination, debonding, bulk fiber pull-out and fiber ruptures were seen commonly in the micrograph results of composite. The EDAX analysis, depicted in Figure 5.8, confirmed the presence of elemental constituents from TSP filler added in kenaf/epoxy composite on the fractured surface of the K5 tensile sample.

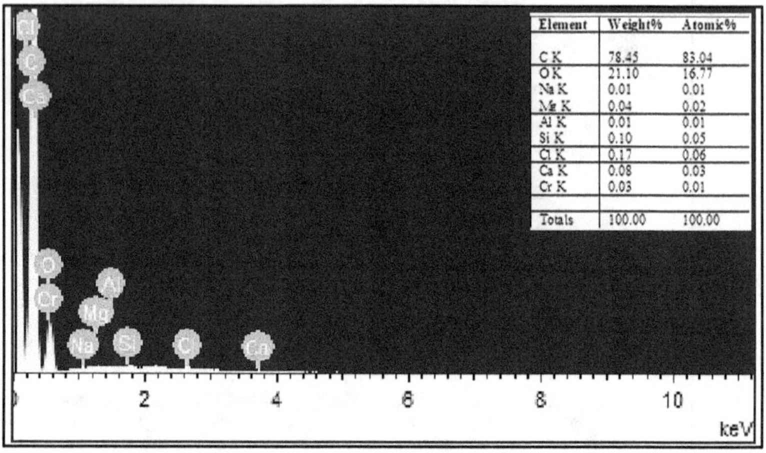

Figure 5.8 EDAX results for TSP/kenaf/epoxy composite of K5 sample.

5.8 Effects of filler and fiber orientation

The presence of TSP filler improves the composite's enrichment, indicating the highest strength of the material as supported by experimental findings. The filler material comprises minerals that consist of microlamellas referred to as keratin. When the minerals are paired for flexibility and toughness mechanisms, this strengthens the composite [Chen et al., (2017)]. The filler improves mechanical characteristics by improving fracture propagation resistance due to the dispersion of biomineralized nanocomposite platelets. The effective micro and nanostructure of the turtle shell filler contribute to the remarkable overall properties obtained. By including this filler, the matrix modifies the bulk characteristics of the composites' interfacial interaction and builds up the nanostructure inside the phase.

The longevity of the kenaf FRP composite has increased stability and decreased matrix failure while reinforcing [Malik et al., (2021)]. The impact of ply direction control enhances the inherent properties, reducing the failure of laminated composites. The symmetrical arrangements of the composite laminates during construction augment their bending stiffness and high impact strength. But, the demerit of kenaf bidirectional fiber is the presence of a gap between the yarns, i.e., pores which allow the matrix to exit from the composite during the fabrication. It reduces the strength of the composite and affects its properties. Moreover, the uniform load distribution is achievable by utilizing an isotropic material composite. However, in this scenario, the quasi-isotropic laminates partially replicate the composite properties [Agarwal et al., (2021)]. The angle control direction of the fibers improves the strain characteristics of each layer when subjected to loads applied parallel or perpendicular to the laminates. It increases the tensile strength in the parallel direction and the bending strength in the perpendicular direction [Salman et al., (2015), Salman et al., (2016)].

CHAPTER 6

SUMMARY AND CONCLUSIONS

6.1 Summary

The development of advanced polymer composites using natural fibers and bio-derived fillers has made significant progress due to the increasing demand for sustainable, lightweight, high-performance structural materials. Basalt and kenaf fibers are two reinforcements widely recognized for their low density, environmental friendliness, and excellent mechanical properties, making them viable alternatives to conventional synthetic fibers. The present study reports the use of Turtle Shell Powder (TSP) as a filler material in two different fiber-reinforced polymer (FRP) systems,

1. Basalt / Epoxy Composite
2. Kenaf / Epoxy Composite

TSP exhibits a unique architecture comprising bio-shielding elements and biomimetic strategies. It naturally protects it from predatorial attacks. Its nanoplatelet structure imparts exceptional stiffness, strength and multifunctional mechanical properties.

6.2 Conclusions

In this work, TSP-reinforced basalt and kenaf FRP composites were fabricated and systematically examined through mechanical characterization, morphological analysis and comparative evaluation.

Characterization of basalt/epoxy laminated composites reinforced with TSP:

- The incorporation of TSP significantly enhanced the mechanical properties, with the 3 wt.% TSP (B3) composite exhibiting the best overall performance.
- The tensile strength, flexural strength, impact strength and hardness of the B3 composite increased by 98.4%, 70.73%, 76.85% and 43.5%, respectively, compared to the neat composite (B0).

- SEM analysis further confirmed improved interfacial bonding between the basalt fibers, epoxy matrix and TSP.
- Bonding enabled better stress transfer and reduced void formation.
- The quasi-isotropic ply orientation contributed to more uniform load distribution and enhanced bending and impact behavior.
- But excess filler loading (5 wt.%) led to particle agglomeration and micro-voids, reducing the mechanical performance relative to B3. It indicates an optimal filler threshold.
- Therefore, the results established TSP as an efficient bio-filler capable of significantly enhancing strength, stiffness and energy absorption in basalt-epoxy composites, highlighting their potential for ballistic, defence and impact-resistant structural applications.

Characterization of kenaf/epoxy laminated composites reinforced with TSP:

- The sample with the 5wt.% TSP (K5) laminated composite demonstrating the highest mechanical performance among all formulations.
- Compared to the control (K0), the tensile strength, flexural strength, impact strength and hardness of the K5 composite improved by 78.06%, 43.64%, 38.7%, and 47.8%, respectively.
- Enhanced dispersion and strong interfacial adhesion among TSP, kenaf fibers and epoxy promoted efficient stress transfer and increased load-carrying capability.
- SEM micrographs confirmed reduced voids, improved fiber–matrix bonding and diminished crack propagation.
- The quasi-isotropic fiber orientation further augmented flexural performance and impact resistance by ensuring balanced load sharing.
- Higher TSP content increased composite density and reduced void content, validating the effectiveness of the filler as a reinforcing agent.
- Henceforward, researchers are encouraging green composite development by combining natural fibers with bio-fillers based on the applications, which is reducing the environmental footprint.

- Therefore, the findings demonstrate that TSP serves as a promising bio-filler capable of significantly enhancing the mechanical performance of natural-fiber composites, making them suitable for applications in automotive, marine, sporting goods and lightweight structural components.

6.3 SCOPE FOR FUTURE RESEARCH WORK

The present research offers substantial opportunities for future investigations and leads to exploring several new dimensions in the development of TSP-based composite materials. Building upon the findings of this study can further broaden the scientific understanding and technological potential of these composites. The major scope for future research includes,

- Investigate filler systems by combining TSP with nanoclay, graphene, CNTs or ceramic particles used in the FRP composites.
- To optimize particle size and chemical treatments of TSP to improve the mechanical strength of the FRP composites.
- To conduct thermomechanical, Dynamic Mechanical Analysis (DMA), Differential Scanning Calorimetey (DSC), creep and fatigue studies for reliability assessments.
- To analysis the performance of high-strain-rate and ballistic test to validate the suitability for defense-grade applications.